The IT in Secondary Science Book

Roger Frost

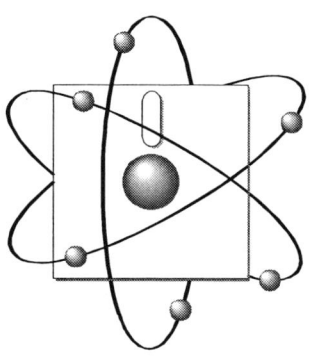

A compendium of ideas for
using computers and teaching science

Titles in this series:

The IT in Science Book of data logging and control - ISBN 0-9520257-1-X
The IT in Secondary Science Book - ISBN 0-9520257-2-8
Data logging in Practice - ISBN 0-9520257-4-4
Software for Science Teaching ISBN 0 9520257 5 2
IT in Primary Science - ISBN 0-9520257-3-6

How to contact the publishers, our suppliers and the author

The IT in Secondary Science Book © was written and produced by Roger Frost
E-mail books@rogerfrost.com. It is published by
IT in Science, 7 Sutton Place, London E9 6EH Telephone or Fax: 020 8986 3526.

For teaching materials and information about IT resources, visit Roger Frost's Dataloggerama on the Internet at www.rogerfrost.com

IT in Science distribution - Please address UK bookshop and bulk purchases enquiries to the publisher.
UK: The Association for Science Education (Booksales), College Lane, Hatfield, Herts. AL10 9AA. Tel: 01707 267411. Fax: 01707 266532 Web: www.ase.org.uk
also Griffin & George Ltd and Scientific & Chemical Supplies (Hogg Scientific)
Australia: Southern Biological Services Pty. Ltd, 19-21 Worrell Street, Nunawading, Vic. 3131. Australia. Telephone: 03 9877 4597 Fax: 03 9894 2309 Email: southernbiological@bigpond.com
New Zealand: Education Advisory Services, Private Bag 92601, Symonds St, Auckland
USA: Fisher Education Tel: 1 800 955 1177 Internet: www.fisheredu.com/
USA: Data Harvest Educational Inc. 349 Lang Blvd, Grand Island, 14072.
Tel: 1 800 436 3062 Internet: www.interlog.com/~ easylog
Canada: Data Harvest Educational Inc. 2671 Romark Mews, Mississauga, Ontario, L5L 2Z4. Tel: 905 828 6166 Fax: 905 607 1525 Internet: www.interlog.com/~ easylog

First published April 1994 and then updated January 1995, April 1996, January 1997, November 1997, January 1998, January 2000 with new releases of software included within the text.

A catalogue record for ISBN 0 9520257 2 8 is available from the British Library. Graphics originated using Arts & Letters Graphic Editor and Micrografx Designer. Printing by Sackville Printing, Piccadilly, London W1

Would you let this man into your school?

When he is not writing and reviewing software, Roger Frost runs computers and science training days for schools and education authorities. He also talks at meetings and conferences on any aspect of using IT in science education. Should you need help or advice in this area, and welcome this sort of person in school, please get in touch.

A list of past and present work can be found on the Internet at www.rogerfrost.com.

Acknowledgements

Thanks are due to the following for their valuable help and advice - in their personal capacities. Their ideas, contributions and comments on the original manuscript were gratefully received and used to prepare this book:

Laurence Rogers, Leicester University School of Education
John Wardle, Microscope IT in Science centre, Sheffield Hallam University
Stuart Robertson, former advisory teacher now at St Lyre's Comprehensive School, Mid-Glamorgan
Dr Angela McFarlane, Senior Lecturer Homerton College IT Unit, Cambridge
Linda Webb, Senior Lecturer, Homerton College
Edited by Rosie Kentish of Deep See Subs and Gerard Killoran, City of Westminster College, London
Cover photograph with thanks to Holland Park School, London

The stimulus for the ideas here came from unpublished and published material including:

IT in Science publications: The IT in Science Book of Data logging and Control (IT in Science)
The IT in Primary Science Book (IT in Science)
IT in Science Buff book (IT in Science)

National Council for Educational Technology publications (now Becta): Practical Science with Microcomputers (NCET)
Science Investigations and IT (NCET)
Enhancing Science with IT (NCET)
Supporting Science (NCET)

Science textbooks: Blackwell Modular Science (Blackwell); Science Scene (Hodder & Stoughton)
Bath Science 5-16 (Nelson); Nelson Science (Nelson), Oxford Science Programme (OUP)
Kaleidoscope (Heinemann);Understanding Science (John Murray) and Active Science.

And also: Creative ideas for using spreadsheets in science by Jane M Morris (Cleveland Education Computing Centre)
Handling Data with Databases and Spreadsheets by Mike Hammond (Hodder & Stoughton)
Electrical Measurements using a computer by Roy Barton (University of East Anglia)
Insight software worksheets by Leicester university (Longman Logotron)
Leicester Science Toolkit worksheets (Deltronics / Leicester university)
Information Technology in Science (MEU Cymru)
Essex Spreadsheet Posters (Essex LEA).

About the author

Roger Frost was a biochemist for ten years before becoming a teacher of science and computing. In 1988 he became a science and IT advisory teacher for ILECC, the London computer centre and later for North London Science Centre. Since 1993 he has worked as a freelance writer, trainer and IT consultant. His published work includes:

The IT in Science Book Of Datalogging And Control (IT in Science) ISBN 0-9520257-1-X
Learning Highways - exploring the potential of the Internet (NCET) Co-author with Roger Blamire
Software for Science Teaching (IT in Science) ISBN 0 9520257 5 2
Data logging in Practice (IT in Science) ISBN 0 9520257 4 4
Enhancing Science with IT (NCET) 1994 Co-author ISBN 1 85379270 5
Science Online (Becta) Ideas for using the World Wide Web in science teaching - 2000 Co-author
IT in Primary Science (IT in Science) ISBN 0-9520257-3-6
Information Technology (Nelson), 1993 Co-author with Roz Reyburn ISBN 0-17-438572-2
The IT in Science Blue book, (IT in Science), 1992 Out of print
The IT in Science Buff book, (IT in Science), 1991 Out of print

Contents

Quick overview

This book is a major catalogue of ideas for teaching science with information technology.

1 Introduction

Introducing information technology in science and its place in the curriculum.

2 Information technology tools and worksheets

This section introduces the different IT tools. It is illustrated with worksheets which show how those tools can be put to work in science.

3 Ideas for using information technology in science

This section lists hundreds of ideas for virtually every part of the science curriculum. Look through the headings to find your topic. See the contents or index for a more detailed list.

4 Reference

This final section lists the suppliers of resources for information technology. It concludes with a glossary of IT tools and an index.

How information technology helps scientists

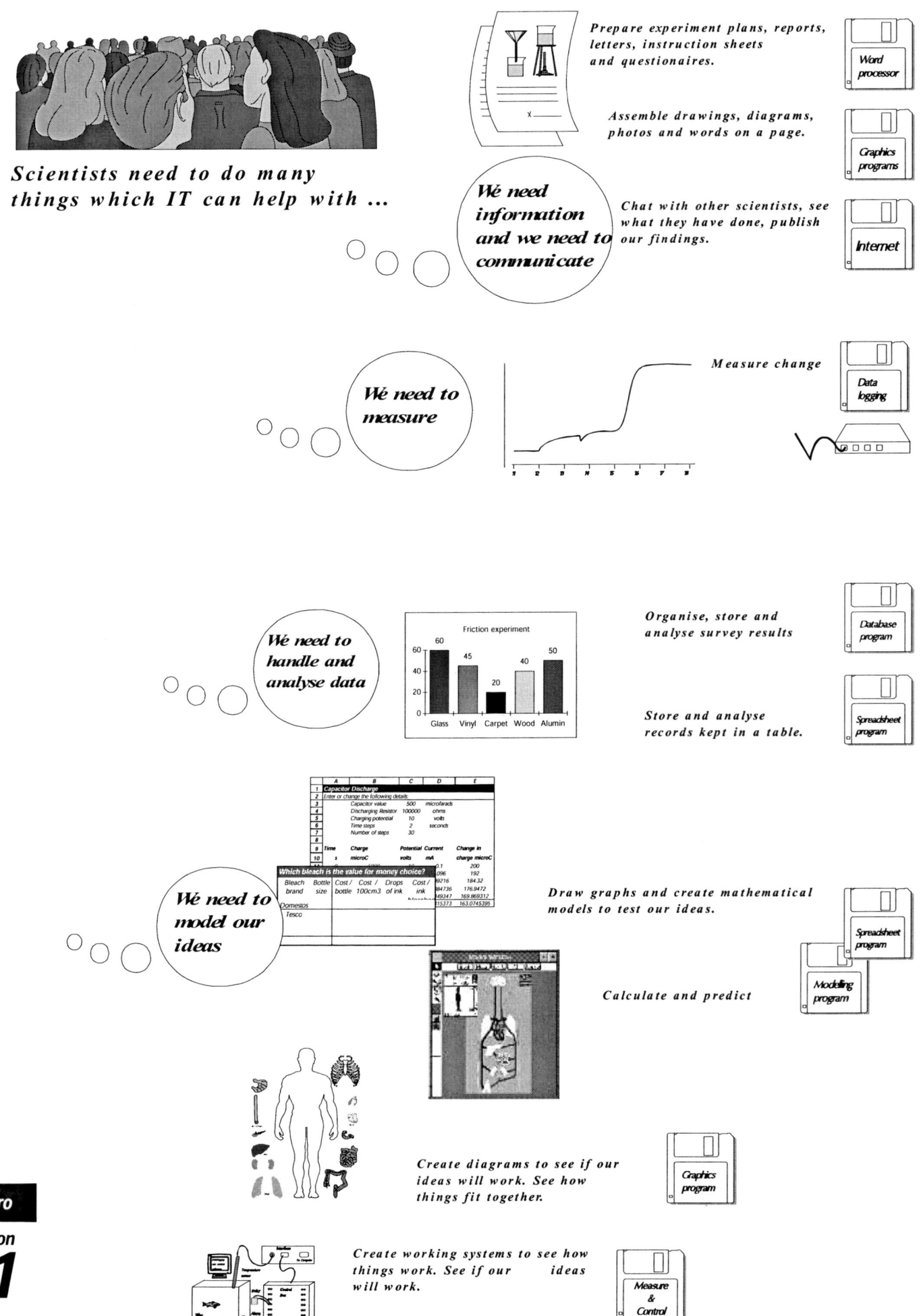

Scientists need to do many things which IT can help with ...

We need information and we need to communicate

Prepare experiment plans, reports, letters, instruction sheets and questionaires.

Word processor

Assemble drawings, diagrams, photos and words on a page.

Graphics programs

Chat with other scientists, see what they have done, publish our findings.

Internet

We need to measure

Measure change

Data logging

We need to handle and analyse data

Friction experiment

Organise, store and analyse survey results

Database program

Store and analyse records kept in a table.

Spreadsheet program

We need to model our ideas

Draw graphs and create mathematical models to test our ideas.

Spreadsheet program

Calculate and predict

Modelling program

Create diagrams to see if our ideas will work. See how things fit together.

Graphics program

Create working systems to see how things work. See if our ideas will work.

Measure & Control

Preface

How information technology helps scientists

Scientists need to measure and communicate, to handle information and model ideas. In essence, they need to process information. Young scientists have similar needs - as they do science work they write, draw graphs, do maths and make measurements - so they too process information.

The technology for processing information includes tools such as the word processor, the spreadsheet, database programs, sensors, and modelling programs. Database programs allow us to search for information and look to patterns within it. Sensors help us to measure changes and draw graphs. Modelling programs help us present scientific ideas that are too hard to get a grip on in real life. Spreadsheet programs take the strain of making tables, drawing graphs and working with numbers. If there is a common thread here, it's that these tools allow us to do more and go further.

It is important that children see how today's scientist works. It is important that they be equipped for the technology-rich world in which they live. It's also important, a legal requirement even, that they use information technology.

But when teachers started using the technology in class, other advantages became apparent. When their pupils became fluent in using sensors, the computer offered a new insight into science: they gained something that helped them to understand and encouraged them to explore. When the children used databases and spreadsheets they didn't just draw graphs, they could go on to interpret them. And when they worked together with a word processor, they started talking with zeal, not the usual gossip, but about science. Children who were challenged by doing things 'the old way' were able to move on. The tools that started life as information processing tools became really special tools to enhance our teaching. These were tools for the mind.

And for all the speed of computers, I doubt if anyone saved any time. What was saved - by not having to draw tables, colour-in graphs, write it out neatly or take thermometer readings - was spent straight away, examining the science that had started to open up. In the search for more science, this book shows where information technology can be exploited and add value to our science teaching.

Roger Frost

Over the years information technology skills have been sorted into several 'strands' or processes:

- *Handling information - which you do when you use database and spreadsheet programs.*

- *Measuring and controlling - which you do when you use sensors and control technology.*

- *Modelling - which you can do when you use spreadsheet and modelling programs.*

- *Communicating with IT - which you do when you use word processors and graphics programs.*

- *The applications and effects of IT - which you can consider as you use information technology.*

While the classification is one of convenience, these strands nevertheless embrace much of the information technology activity that takes place in school.

The tables on these pages show the sort of progress that pupils might make through each of those strands. More importantly, the tables show how science activities using IT can become increasingly challenging.

Handling information

Progression in handling information with information technology	What the pupils do in science	IT level
Explore information held on IT systems.	Go to the computer and find out something about ... snails.	Level 1
Use IT to sort and classify information and to present their findings.	Look at a CD-ROM on animals and show what you found out.	Level 2
Use IT to save data and access stored information, following straightforward lines of enquiry.	Look at a CD-ROM on animals and show us all the animals that live in the jungle. Keep a record of what you found out.	Level 3
Can add to, amend and interrogate stored information. They understand the need for care in framing questions when collecting, accessing and interrogating information. Interpret their findings, question plausibility and recognise that poor quality information yields unreliable results.	Use *BodyMapper* software to add your own details to a class database: add your height, hair colour and so on. Check the details and sort the children into order of height. Make a list of all the children with brown hair.	Level 4
Select the information needed for different purposes, check its accuracy and organize and prepare it in a form suitable for processing using IT.	Do a survey of people in your class. Collect the data and create a class database. Check the data you entered and correct any errors. Search, sort and graph the data.	Level 5
Use complex lines of enquiry to test hypotheses.	Get a database on the chemical elements and use it to find groups of elements and look for patterns.	Level 6
Identify the advantages and limitations of different data handling applications, and select and use suitable information systems, translating enquiries expressed in ordinary language into forms required by the system.	Compare a simple text database of the chemical elements with its CD-ROM equivalent and comment on the strengths and weaknesses of each.	Level 7

1

Assessing information technology

Measuring and controlling things

Progression in measuring and controlling things	What the pupils do in science	IT level
Recognise that everyday devices respond to signals and commands and they can make them respond in different ways.	Talk about how to use a video recorder.	Level 1
Control devices purposefully and describe the effects of their actions.	Technology: introduce robots.	Level 2
Understand how to control equipment to achieve specific outcomes by giving a series of instructions.	Technology: control a robot.	Level 3
Use IT to control events in a predetermined manner, to collect physical data and display it.	Technology: control a robot and make it perform a set routine. Use sensors to make measurements and display readings.	Level 4
Create sets of instructions to control events, and become sensitive to the need for precision in framing and sequencing instructions.	Technology: control a robot and make it perform a set routine.	Level 5
Develop, trial and refine sets of instructions to control events, demonstrating an awareness of the notions of efficiency and economy in framing these instructions. Understand how IT devices can be used to monitor and measure external events, using sensors.	Technology: control a robot, make it perform a set routine and not be content with just getting it to work. Use sensors to make measurements, for example, use digital sensors to measure their reaction time.	Level 6
Use IT equipment and software to measure and record physical variables.	Use sensors to make measurements in experiments. Use a data logger to record the room temperature and light level over a weekend. Display readings as time graphs.	Level 7
Select the appropriate IT facilities for specific tasks, taking into account ease of use and suitability for purpose. Design successful means of capturing and preparing information for computer processing. When assembling devices that respond to data from sensors, they describe how feedback might improve the performance of the system.	Use sensors to make measurements in experiments. Select appropriate sensors and recording parameters. Use the data in the data logging program or export it to a spreadsheet or word processor. Develop a control system to run a biofermenter, an aquarium or fire alarm. Discuss and document the task to a high standard.	Level 8

Modelling

Progression in using computer models and simulations	What the pupils do in science	IT level
Use IT-based models or simulations to investigate options as they explore aspects of real and imaginary situations.	Use educational games on the level of *Zoo Keeper* where pupils have to feed different animals the correct food.	Level 2
Use IT-based models or simulations to help them make decisions and are aware of the consequences of their choices.	Use *Badger Trails* (Sherston) to navigate a badger home past hazards such as the road or lack of food.	Level 3
Use IT-based models or simulations to explore patterns and relationships, and make simple predictions about the consequences of their decision making.	Use a spreadsheet to calculate braking distances of a car under wet and dry conditions. Use the program *Moving Molecules* to model kinetic theory. Or use the program *Botanical Gardens* to study seed growth. Use a spreadsheet program to study data on the planets. Look for patterns in the data. Suggest ideas such as 'the temperatures on the planets increase as they get closer to the sun' and test them.	Level 4
Explore the effects of changing the variables in a computer model.	Experiment with a model of the use electricity in the home. Experiment with a model showing your daily requirements for energy.	Level 5
Use computer models of increasing complexity, vary the rules within them, and assess the validity of these models by comparing their behaviour with other data.	Use a model such as the *Creatures* population simulator (Future Skill).	Level 6
Design computer models or procedures, with variables, which meet identified needs.	Build a spreadsheet to model heat loss from the home. Use it to find the most cost-effective methods of home insulation.	Level 7

1

Assessing information technology

Communicating with IT

Progression in communicating with information technology	What the pupils do in science	IT level
Use IT to assemble text and symbols to help them communicate ideas.	Prepare a captioned picture using a word processor.	Level 1
Use IT to help them generate and communicate ideas in different forms, such as text, tables, pictures and sound. With some support, they retrieve and store work.	Prepare a poster using a word processor - both writing and adding pictures to it. Come back to it and finish it later.	Level 2
Use IT to generate, amend, organize and present ideas.	As above, but using the computer to improve on work they have already done.	Level 3
Use IT to combine different forms of information, and show an awareness of their audience.	Use a word processor to prepare a report of an experiment for public consumption.	Level 4
Use IT to organize, refine and present information in different forms and styles for specific purposes and audiences.	Take a report from an experiment and re-organise it to make an information leaflet.	Level 5
Develop and refine work using information from a range of sources, and demonstrating a clear sense of their audience and purpose in their presentation.	Make an advertisement for Aluminium Metal using words and graphics.	Level 6
Combine a variety of forms of information for presentation to an unfamiliar and critical audience.	Choose and use software for a poster, newsletter or multimedia presentation	Level 7

Applications and effects of IT

Progression with the applications and effects of information technology	What the pupils do in science	IT level
Describe their use of IT, and its use in the outside world.	Discuss how scientists or others might use the IT tools they use.	Level 3
Compare their use of IT with other methods.	Each time pupils use a new IT tool, discuss its advantages and disadvantages. For example, say how a class database compares with a class register. Or when pupils display graphs on the computer, compare these with hand drawn efforts.	Level 4
Communicate their knowledge and experience of using IT and assess its use in their working practices.	Pupils explain how sensors help them do their experiments.	Level 5
Discuss the wider impact of IT on society.	Talk about the applications of sensors in everyday life, about the growth of technologies such as interactive television and the information superhighway.	Level 6
Consider the limitations of IT tools and information systems, and of the results they produce.	Pupils come to appreciate the things they can and can't do using sensors and data logging software.	Level 7
Discuss in an informed way, the social, economic ethical and moral issues raised by IT.	Discuss issues such as: rapid obsolescence of IT; RSI; computer games; effects on jobs and even the positive aspects of IT.	Level 8

1

*T*oday's scientists need **database programs** to handle their data. They also need to read about other scientists' work, to find data and analyse it. They may search through a database on disc, on the Internet or even on a CD-ROM. They can use their science skills to collect, organise and analyse data. They think carefully and look for patterns, they think critically and check data for errors. They will see how their findings fit other people's. Using a database in school parallels these processes.

Database programs allow you to store, sort and graph the results of a survey or investigation. If you had a database of pupil's personal data, you could sort the pupils into order of shoe size, or work out the average for the class. You could draw a bar chart to see how the shoe sizes vary across the class. Or draw a scattergraph to see if they vary with height. You might also search for all those with black hair and see if they have any eye colour in common. Using a database provides many opportunities to analyse data. It is a great tool for exploring science.

There's an good number of ready-made databases waiting to be analysed. On the Internet you'll find data on world health and world energy use. Or you can buy data files about minibeasts, the elements, the planets, mammals, birds, plants and rocks. Sometimes, as with CD-ROM data, the data is enriched with photographs, animation and sound which makes things accessible to younger pupils.

Pupils can also learn to make their own databases. They can study themselves or survey the world at large. Either way, when they make a database they have to be quite scientific in how they work. They have to define what they want to find out, collect the data, organise it and ultimately analyse it. All of this belongs, and deserves a place in science teaching.

IT tools

Section
2

Database glossary

Alphanumeric field - a type of field where you can store a mixture of numbers and letters.

Choice field - a type of field which gives you a list of items to choose from. A 'choice field' for eye colour could list blue, black, brown and green while, in a weather database, a choice field might offer dull, bright, cloudy and so on. Using this type of field can not only prevent errors and inconsistencies when you enter your data, it also encourages children to classify the objects in their database. When you use a database program in science, look for this feature in particular.

Database - a collection of data about one topic. In school practice a database is a file of records. In commercial practice a database can be a massive set of files about one topic.

Field length - tells the computer how much space an item of data takes up. The length of the field is measured in characters or key presses, i.e. the name red robin takes up 9 spaces, but you need extra space for longer names such as woodpecker.

Field types - different types of data need different field types. The most common field types are number and word types. Numbers often need units such as cm and kg.

Field names - often referred to as headings. These are headings such as height, eye colour and shoe size. A **field** is the part of a file which stores a piece of information. The headings above might have fields storing 120cm, blue and 6.

File - a collection of records. In practice it is what you save on the disc.

Chart - a feature to draw pie, bar and line graphs with the data. See the **Graphing tools glossary** on the next page.

Numeric field - a type of field where you can store numbers only. You would need a numeric field to store heights and weights.

Pictogram - a symbolic bar chart. The symbols can be coloured blocks or pictures such as a stack of cars, aeroplanes or other symbol.

Records - a set of fields about one thing is called a record. The form you fill in on the screen about one thing or person is called a record.

Scattergram - a table which shows a pattern between two fields - e.g. you can show whether blonde haired children have blue eyes. A scattergraph shows you a pattern between two numeric fields. A scattergraph plots the fields as a series of dots and as you look along the dots, a line through them indicates a pattern.

Search - lets you select out certain records - e.g. search for all the people with brown eyes. You can do 'complex' searches where you search for all the people with brown eyes and black hair. You can use a search in the hunt for patterns.

Spreadsheet - another type of program which can handle data. These programs share many features in common with databases. If your set of data is quite small, say, just 10-20 records and you just want to draw graphs, a spreadsheet may be a better choice.

Sort - puts all the records into either alphabetical or numerical order. The direction of a sort can be in ascending or descending order.

Database jargon

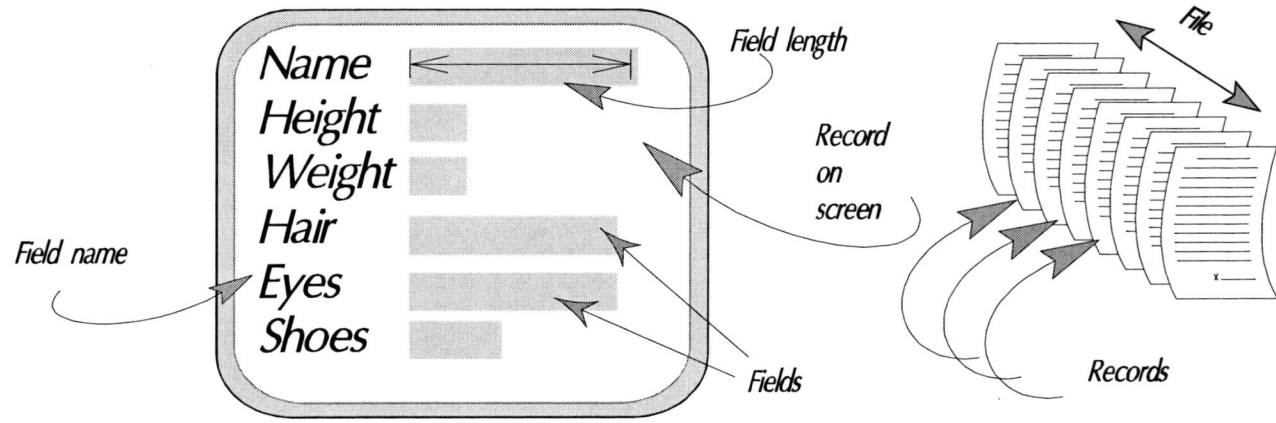

Graphing tools glossary

Graphs are a key tool for analysing data and computers draw them with great ease. In fact, when you use databases and spreadsheets you can produce an astounding range of graphs. Our role as science teachers could therefore be to encourage pupils to communicate effectively using graphs.

Here then are some working descriptions of the most popular and useful kinds of graph you will meet on the computer.

Histograms and count graphs

These give an idea of the distribution of your results. For the chart here, the complete range of insect lengths were divided into five equal ranges and counted. The graph shows the number of insects which have similar lengths. Histograms, unlike bar charts, show which ranges are the most significant and whether the results are well distributed or skewed. Some programs let you make pictograms - showing pictures instead of bars.

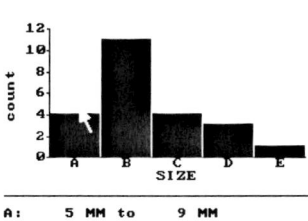

Pie charts

One of the easiest charts for comparing parts with a total. For example, you can plot a pie chart showing the different gases in air or as in the example here, you can see what proportion of a class are girls. When you use a **database** a pie chart might show you, for example, the spread of the shoe sizes in the class. (For obscure reasons, a pie chart with the same data in a spreadsheet may not).

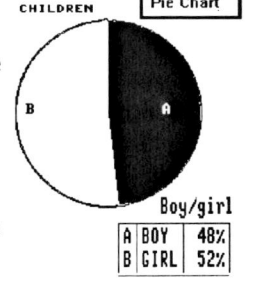

Bar or column chart

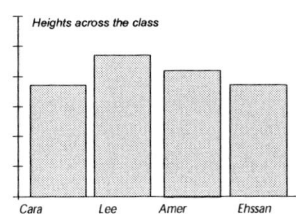

The term bar chart is a generic one. There are stacked bar charts, histograms and more. In computer usage a bar chart shows the spread of the results. For example, a bar chart of pupils' heights shows each pupil with a bar representing their height. A histogram of the same data would divide the class into ranges and count their number falling in each range.

Scattergraphs or X-Y graphs

The most useful graph for science. These help to find a pattern between two sets of numbers or variables - for example, to find out if larger animals have larger wingspans or see how current changes with resistance. Usually **(see below)** you see a pattern of dots - rather than a line of best fit. However, while computers can draw the best-fit line for the pupils, many teachers feel that the pupils should draw it for themselves.

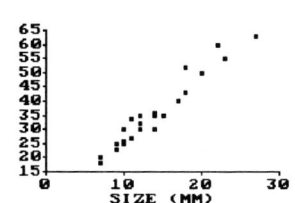

Venn diagrams

Useful for seeing if there is a connection between different features, for example, do minibeasts with two wings feed on nectar? The circles show how many creatures have each feature, while the overlap shows how many have both features.

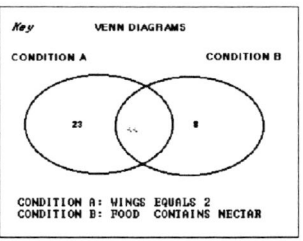

Line graphs

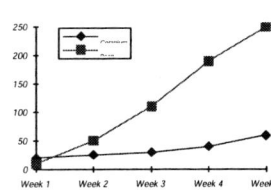

Computers, like pupils, think that line graphs are just bar graphs drawn with a line instead of bars. For example, you can use a computer line graph to show plant growth over time. (However, you must ensure that your readings were taken at equal intervals - be that days or weeks). Essentially, line graphs have similar uses to bar graphs. If, in fact, you really want a graph where one set of numbers is plotted against another, ask for a scattergraph instead.

Database ideas

Making a database about your class

This first page is a teacher's guide for a database project. The five sheets which follow it form part of this project. These are:

1. Collecting data for a database - page 18

This asks the pupils to be clear about the questions they want their completed database to answer.
IT level: easy

2. About computer databases - page 19

This sheet is an IT information sheet and paper exercise focusing on database terminology and designing a database structure.
IT level: easy/medium. (The test questions are not so easy).

3. Designing a database for your survey

This sheet, on p. 20 is a paper exercise on designing a structure for a database.
IT level: easy/.

4. Analysing data I - page 21

This is a computer exercise on using a database to analyse data. It looks at the choice of graphs available and if pupils have a data file they can try the ideas on the sheet with it.
IT level: easy/medium

5. Analysing data II - page 22

This too is a computer exercise on using a database to analyse data. It looks at averages, sorting and drawing bar graphs and scattergraphs. With these tools they can spot trends and patterns. If pupils have a data file they can try the ideas on the sheet with their own file.
IT level: easy/medium.

Minibeast hunt - page 23

This is a simple activity for learning-to-search with a database. The aim is to get the difficult idea of using search after search to narrow down the options. For example, if there were 150 animals on file, one search might narrow that down to 20, another search might narrow that down to 6, another to one or two. The activity has built in 'success' in that the pupils finally see a graphic of the animal they seek. It requires the **Anglia TV Key** data file on **Minibeasts of Britain and Ireland**. There are now 'minibeast' data files on CD-ROM which make things a little more exciting and realistic.
IT level: easy.

Minibeasts II - page 24

An activity for learning-to-search and draw graphs with a database. As above, this activity was built around the same very detailed **Key** data file on minibeasts. The bank of questions can be adapted for other files.
IT level: medium.

The planets - page 25

An activity for drawing graphs and looking for patterns with a database. You can buy ready-made files although you can quickly make your own file using the table of data supplied.
IT level: medium.

The weather - page 26

A question bank for analysing data from a weather station. You can use your classroom sensors to collect the data or indeed use an automatic weather station. The first approach requires good IT skills to get the data in the computer. The latter approach is more realistic.
IT level: medium.

The chemical elements - page 27

A question bank for analysing data from a file on the chemical elements. You will find ready-made databases on the elements on disc and CD-ROM. There are also many programs on the elements which provide the necessary features to do this. See the main section under **Materials**.
IT level: medium/hard.

Some more ideas for database projects

Human variation - classes and whole year groups.
Food - which foods have the most energy?
Fruits - how are they similar and different?
Sycamore wings - do those with wider wings fall slower? Do those with longer wings fall faster?
Cylinders - do larger diameters roll further down a slope? Do heavier tubes roll further?
Paper aeroplanes - what makes a good paper aeroplane? How far should the wings be from the front? Should they have weights on the tail?
Other database projects - musical instruments, birds, animals, leaves, creatures in leaf litter, household objects, liquids, household chemicals, materials, metals, seeds, beans, cars, waste and rubbish, rocks and minerals.

IT tools

Section
2

Making a database about your class

A database project positively needs planning. While the following pages are for pupils, this page is a walk-through 'teaching scheme' for a class database project.

Starting from the basics

If the pupils have no prior experience of databases you can try the following as a preamble. Make a card-index database on the class. You might sort the cards in order of height and name. Or you might arrange them as a bar chart of heights on the desk (cf. diagram above). You can pick out those with black hair and similarly arrange them as a bar chart on the desk.

Analyse a database of pupil data.

Sort, search and graph a ready-made database. See the two sheets that follow on *Drawing graphs* which show pupils various ways to find things out with a computer.

Add your own entry to the database.

Pupils can add their own personal data to an existing database. They might then draw a bar chart, say of pupil heights and see where they themselves appear. Following this dry-run they can set about preparing a database for real.

Decide what you want to find out.

You might want to know: who is the tallest in the class? Who is the oldest? Does arm reach have anything to do with height? Which eye colour is the most common? What are our favourite foods? See the sheet on *Collecting data for a database project.* This easy sheet looks at this important part of doing a survey.

Decide on the data you need to collect.

For example you might collect details of height, birthdate, shoe size, arm reach, eye colour, hair colour and favourite food. See the *Using IT* section on *Genetics and variation* for many more ideas.

Decide on how you will collect the data.

List all of the items of data you need to collect on a data collection form or questionnaire. Also say which units to use, for example, height in cm, gender as B or G, birthdate as YYMMDD, shoe size as Euro or UK, arm reach in cm, reaction 'time' in cm or s, eye colour - choice of blue, black, brown or green, favourite food as choice of beans (not baked beans), chips (not fries). Print off the questionnaires and collect the data.

Get the information ready and check it for errors.

Check the completed data collection sheets for errors and inconsistencies.

Create the heading (or field) names.

Depending on your software, you may also need to shorten long headings. For example, 'hair colour' might have to be shortened to 'hair'. See the two sheets that follow called *About computer databases* and *Designing a database for your survey.*

Enter the data.

The pupils can now enter their own data. Saving their work regularly guards against them losing it.

Check the information in the computer.

Print out the data and check it away from the computer.

Become familiar with the data file.

Refer to the questions that you started the project with. Use the database program features which allow you to sort, graph and average to answer them. See the two sheets that follow on *Drawing graphs* which put this into practice.

Evaluate the work

Younger pupils may find this difficult. See the section on *Assessment - applications and effects of IT* for some of the points you might raise.

IT tools

Section
2

Collecting data for a database project

What this is about

You have been asked to do a database project to find out how much the people in your year group vary.

Before you start collecting data about everyone you obviously need to know what you want to find out.

Your database has to answer the following questions:

What is the average height of your group?

How many pupils are taller than average?

Are there more pupils who are shorter than average?

What is the most common hair colour?

What is the most common shoe size?

Do taller people take bigger steps?

Do shorter people have wider shoulders?

Are younger people shorter?

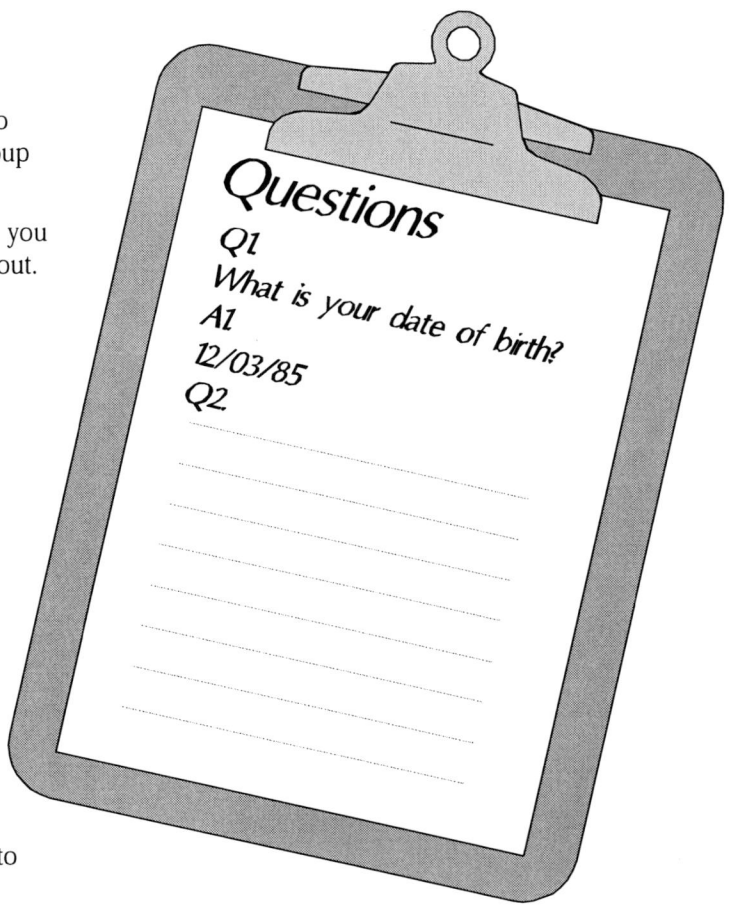

What to do

Write down two more questions you might try to answer.

I might also try to answer these questions:

You now need to collect some data from the group. Write a list of the things you need to collect.

I need to collect data on
Height

Make a data capture form or questionnaire which you can use to do your survey.

Extra tip

If you use a word processor you can not only write your questions, you can also test your questionnaire out on a friend and if necessary change your questions.

About computer databases

What this is about

How database programs use some unusual terms to describe how they organise information.

About organising information

We wanted to study how much we, as humans vary. We organised our information under different headings ready to store in the computer.

There was a heading for the name, a heading for how tall the people were, a heading for their shoe size, their hair colour, their eye colour and their birth date. In computer-speak, the headings are called **field names**.

We entered our information, in spaces next to the headings - these spaces are called the **fields**.

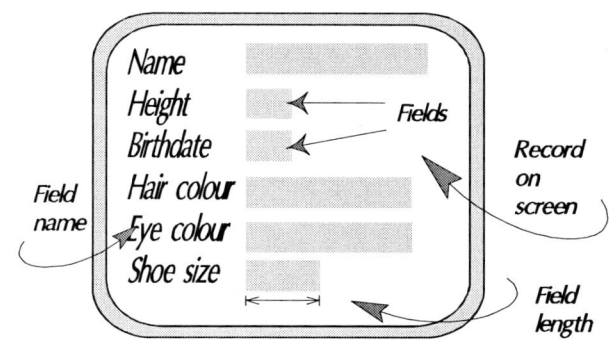

Database terms: fields, field names and field lengths.

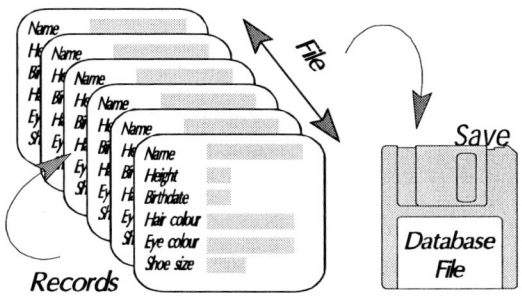

Database terms: records and file

Soon, we had entered information about each person. All the information about one person is called **a record**. We saved the records on a disk. All the records together make what computers call **a file**. (See above)

What type of field?

The fields store information. This might be words, numbers, dates or choices.

The name field is **alphanumeric** - meaning that it can contains words, letters or numbers. The **name** 'Jo' is alphanumeric.

The **height** field is **numeric** - it contains numbers only. A **height** of '120' is numeric.

The **birthdate** field is a **date** field. '6/6/82' is a date.

The **hair colour** is a **choice** field - to force you to choose from brown, black, blonde or red when you enter your information.

Some pieces of information take up more space than others. We call this space the **field length**. For example, the field for **hair colour, needs** 6 characters to store the colour 'blonde'.

Why do all this?

Once you have the information in the computer:

You can sort it into different kinds of order. So you might sort the class into order of their height.

You can draw a chart. A pie chart could show you the spread of hair colour across the class. A bar chart could show you how the heights of the class vary. An x-y graph can show you if taller people have bigger feet.

You can search the information to find all those with black hair. You can go on to find all those with black hair AND brown eyes. You can go further and ask "are people with black hair and brown eyes taller than average".

Test yourself

Suppose you wanted to study all the different liquids you can buy at the supermarket. You want to find out what most liquids are used for, what they cost, the size of a bottle, whether they contain colouring and whether they are poisonous, corrosive or edible.

1. What information do you need to collect?

2. Which field headings will you use?

3. Which fields might be choice fields?

4. Which fields have numeric information?

5. Which fields have alphanumeric information?

6. Which field might take up the most space?

7. Write out one of your records.

8. What would you do to show the range of sizes of the bottles?

9. What would you do to show the range of colours of the liquids?

10. What would you do to show whether most of the liquids are poisonous?

IT tools

Section

2

Designing a database for your survey

What this is about

Here you can practise the design of a database to store your survey results.

What you did

Suppose you had just done a survey of the people in your year group. You now want to design a database to store the information you have collected.

For example, we collected the following data. The table shows you six **fields** belonging to one **record** in our database.

Data collected	For example
Name	Jo Smiff
Date of birth	12.3.85
Height	110 cm
Hair colours	Black
Shoe size	6 cm
Shoulder size	60 cm

What to do

Design your database by filling in the diagram below. Either use your own survey or use our example if you wish:

Write the field names

Write in the field names or headings you will use

Fill in one record

Write in one record from your survey.

Write the field types

Write in the types of field you will use: some will be numeric, some alphanumeric and some might be choice fields.

Mark the field lengths

Write in how long each field should be - i.e. how many key presses or character spaces, at most, will you need?

Write the field names here   Fill in one record here   Write the field types here   Mark the field lengths here

Eye colour   Brown   Choice   6

Field name   Example data   Field type   Field length

IT tools

Section

2

Analysing data I

What this is about

This page shows how you can use a computer to analyse your data. This involves choosing the best type of chart to draw.

If you have a database about a group of pupils, you might use this page to analyse it.

Choosing the best graph

This is the data about a class of pupils. Across the top are the field names or headings.

Name	Colour of Eyes	Colour of Hair	Height cm	Weight cm	Shoe Size	Boy/Girl
Sertac	Brown	Black	129	29	3	Boy
Geoffrey	Blue	Brown	131	30	3	Boy
Ehssan	Brown	Black	131	32	3	Boy
Yit	Brown	Brown	138	32	4	Boy
Sonia	Blue	Brown	130	33	3	Girl
Tony	Hazel	Blond	133	34	2	Boy
Cara	Blue	Blond	137	37	5	Girl
Alistair	Brown	Black	132	37	4	Boy
Sam. A	Brown	Black	143	38	5	Boy
Sam. M	Brown	Black	137	39	5	Boy
Nahum	Brown	Black	142	40	5	Boy
Mustafa	Brown	Black	137	41	3	Boy
Paula	Brown	Brown	144	42	5	Girl
Andy	Brown	Black	143	42	5	Boy
Lee	Blue	Blond	142	43	4	Boy
Derek	Brown	Brown	145	43	4	Boy
Yucel	Blue	Brown	147	44	5	Boy
Amer	Brown	Black	144	47	5	Boy
Victor	Brown	Black	151	51	7	Boy

One way of using the computer to find things out is to draw graphs. For example, we wanted to find out about the mixture of boys and girls - so we drew a pie chart using the Boy/Girl field:

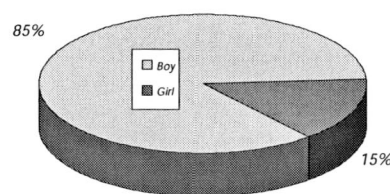

Boys and girls in our class

What does this chart tell you?

We wanted to find out about the class' shoe sizes so we drew a pie chart of the Shoe size:

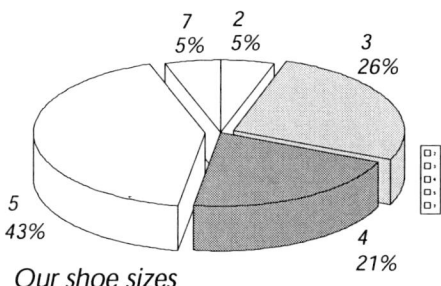

Our shoe sizes

What are the two most common shoe sizes?
What are the two least common shoe sizes?

We also drew a bar chart of the Shoe size:

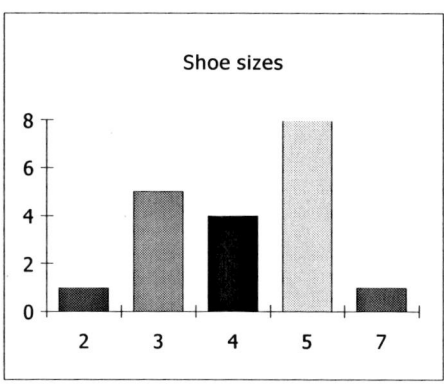

Then we drew another type of bar chart:

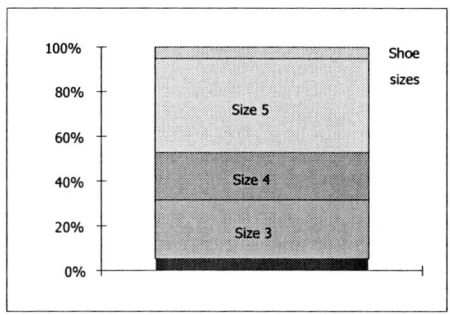

Questions

Which of the two bar charts best shows the spread of shoe sizes across the class?

Compare your favourite **bar chart** with the **pie chart** above. Which best shows the spread of shoe sizes across the class?

How could you find the most common hair colours in the class?

How could you find the most common eye colours in the class?

IT tools

Section
2

Analysing data II

What this is about

This page shows how you can use a computer to analyse your data. This involves sorting lists, working out averages and drawing charts.

Showing how the results vary

We collected some data about a class of pupils. There was data on height, weight and shoe size. We worked out the average weight of a pupil and then set about finding how everyone's weights varied by drawing a bar chart:

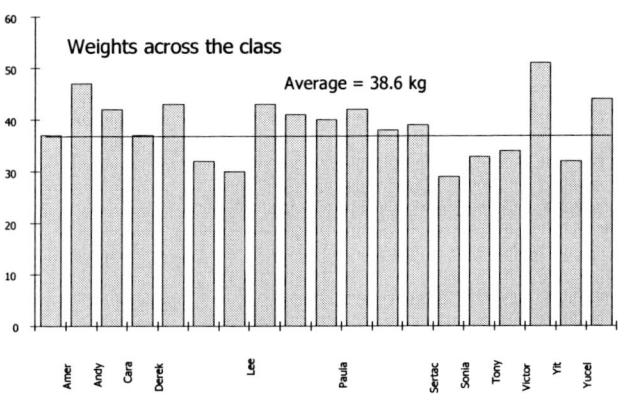

However, this was not very useful. We cannot easily see how many are above or below the average weight. We therefore decided to sort the class in order of weight and draw the bar chart again:

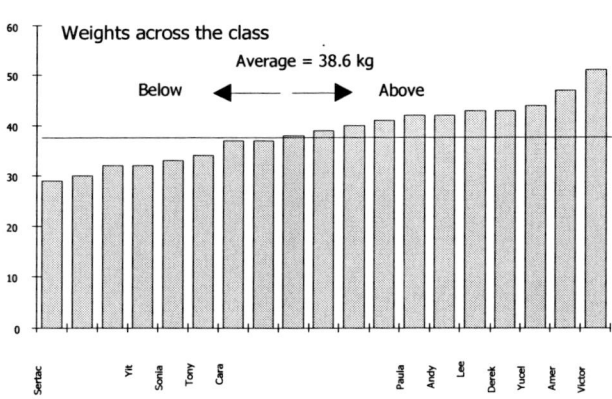

How many pupils are above the average weight?
How many pupils are below the average weight?

Looking for patterns

You can look for patterns by drawing a scattergraph. We wondered if there was a pattern between height and weight so we made a scattergraph of height against weight:

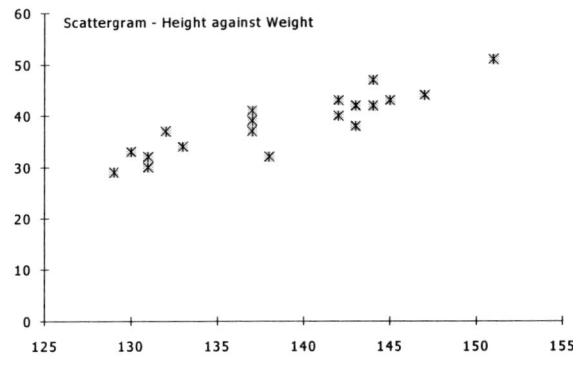

If you can see a trend in the points you may have found a pattern. What is the pattern in the chart above?

Showing the most common results

We wanted to find the most common weights in the class so we drew a histogram. This is a type of bar chart where all the data is arranged in groups. This is a histogram of the class' weights:

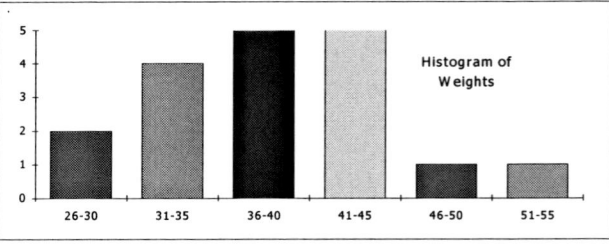

What are the most common weight groups in the class?

Using your own database

Use your database to find answers to the following:

How many pupils are above the average height?
How many pupils are below the average height?
(Find the average and sort the list. Draw a bar chart)

Do taller pupils wear bigger shoes?
(Draw a scattergraph of shoe size against height).

Do heavier pupils wear bigger shoes?
(Draw a scattergraph of shoe size against weight).

What are the most common height groups in the class?
(Draw a histogram of the pupils' heights).

Minibeast hunt

What this is about

In this activity you practice hunting for minibeasts hidden in a computer database.

Suppose you had just found one of the following creatures, could you identify it? A computer database can store information about these creatures and help you to find out more about them.

What to do

Get your computer database on minibeasts on the screen. Search the database to find and identify the creatures we found. We have also provided some clues as to what you should do.

What is this?

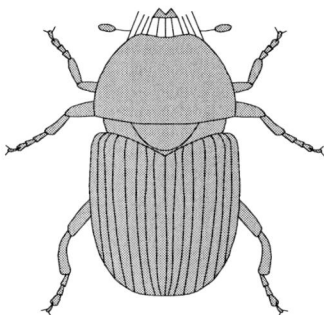

We saw this creature under a log. It has a red body and 18 legs.
Start a new search.
Search on legs equal to 18.
Search again on Habitat.
Then search again on Colour.

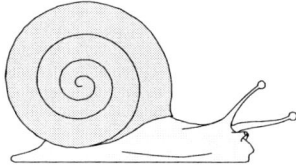

We found this creature on the garden compost heap. It has a grey body and a brown shell.
Start a new search.
Search on legs equal to one.
Search again on Colour.
Then search again on Shell colour.

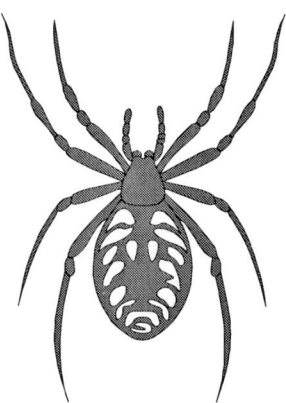

We found this minibeast too. It has a green body.
Start a new search.
Search on Wings equal to ???
Search again on ???
Then search again on ???

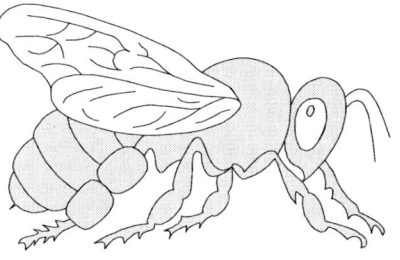

We saw this in some flowers.
Start a new search.
Search on ???
Search again on ???
Then search again on ???

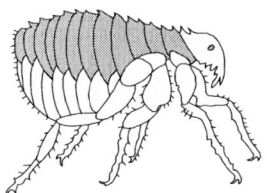

This creature moves fast. It jumps and has a hard, brown body.

Start a new search.
Search on Body/Wing size less than ???
Search again on ???
Then search again on ???

Note:

This activity was built around the **Anglia TV Key** data file on **Minibeasts of Britain and Ireland**.

IT tools

Section
2

Minibeasts II

What this is about

In this activity you use a computer database to look for patterns in the world of minibeasts.

You can use a database about minibeasts to find information and patterns.

Questions

1. To what families (or classes) do minibeasts belong?
   Draw a pie chart of the minibeast Families or Classes

2. How many legs do insects have?
   Search for all the insects.
   Draw a pie chart of how many legs they have.

3. Are insects always black in colour?
   Search for all the insects.
   Draw a pie chart of their colour.

4. How many wings do insects have? Are there any insects with no wings at all?
   Search for all the insects.
   Draw a pie chart of their wings.
   Complete this: Most insects have _____ wings.
   The _____ has/have no wings.

5. What kind of eyes do insects have?
   Search for all the insects.
   Draw a pie chart of their eyes.

6. How do insects breathe?
   Search for all the insects.
   Draw a pie chart of their breathing method or oxygen uptake.

7. Which insect/s have simple eyes instead of compound eyes?
   Search for all the insects with simple eyes. Complete this: The _____ has simple eyes instead of compound eyes.

8. Which type of insect has only two wings?
   Start a new search. Search for all the insects with wings the same as 2.
   Browse through the insects to see what they have in common, then complete this:
   The insects that have only two wings are different kinds of _____ .

9. What do two-winged insects eat?
   Search for all the insects with wings the same as 2.
   Browse through the insects. Draw a pie chart of their food. Complete this:
   Insects with two wings, such as the _____ eat foods such as ____ , _____ and _____ .

10. What do four-winged insects eat?
    Start a new search. Search for all the insects with wings the same as 4. Draw a pie chart of their food.
    Complete this: Insects with four wings, such as the _____ feed mostly on ____ .

Finding patterns in your data

When you need to show patterns in your data a Venn diagram can help you.

What is the pattern between the number of wings an insect has and whether it feeds on nectar?
Start a new search. Search for all the insects.
Draw a Venn diagram (3 sets) to include those with 2 wings, those with 4 wings and those that feed on nectar.
Complete this:
Most insects with _____ wings live on nectar. Most/all insects with _____ wings do not.

True or False

Some of the following statements are false. Use your database to find out. Copy the statements and say whether they are true or false.

All insects have compound eyes. *Search for insects and draw a pie chart of their 'eyes'.*

Some spiders can fly. *Search for spiders and draw a pie chart of the number of wings they have.*

Insects are not nocturnal (active at night). *Search for insects and draw a pie chart of their 'activity'.*

Woodlice are crustaceans. *Search for woodlice and draw a pie chart of their family or class.*

Woodlice like fish, have gills. *Search for woodlice and draw a pie chart of their method of 'oxygen uptake'.*

All worms are hermaphrodite (both sexes). *Search for worms and draw a pie chart of how they reproduce.*

All slugs and snails have a shell. *Search for shell and look at the records.*

The planets

What this is about

You will use a database of the planets to find information and patterns.

Finding out about the planets

Search your database to find:

The planets with a day longer than earth's.

The planets with a day shorter than earth's

The planet with a year which is less than its day.

Sorting

Use the Sort command to list the planets in order of their:

- Distance from the sun

Which planets are earth's nearest neighbours?

- Gravity

On which planet would you weigh the most?

Bar charts

Draw a bar chart of the planets' distances from the sun.

Which is closer to the earth, the sun or Jupiter?

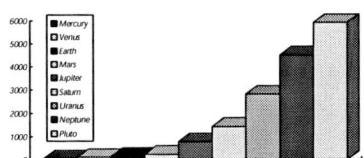

Draw a bar chart of the planets' densities.

Which planets have a density similar to earth's?

Draw a bar chart of the planets' diameters.

Which are the two largest planets?

Finding patterns

You can use scattergraphs to look for patterns in this database. For example, we asked if there was a pattern between the mass of the planets and their diameter. As you would expect, the scattergraph shows that the heavier planets are also the larger planets.

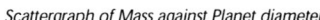

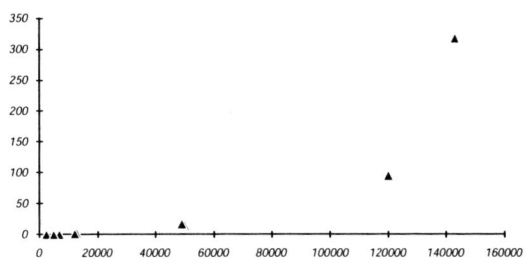

Scattergraph of Mass against Planet diameter

There are more patterns to find and explain. To answer each of the following, draw a scattergraph, see if there is a pattern and try to explain it:

Is there a pattern between the density of the planets and their distance from the sun?

Is there a pattern between the mass of the planets and their distance from the sun?

Is there a pattern between the density of the planets and their mass?

Is there a pattern between the length of a day on a planet and its mass?

Is there a pattern between the length of a day on a planet and its diameter?

Is there a pattern between the average temperature of the planets and their distance from the sun?

Is there a pattern between the time planets take to go round the sun and their distance from the sun?

Is there a pattern between the length of a day on a planet and its mass?

Fieldname	Diameter	Distance	Mass	Temperature	Moons	Orbit time	Gravity	Day	Density
Units	km	million km	Earth=1	deg C		Earth years	Earth=1	hours	tonnes/m3
Jupiter	143000	780	318	-120	14	12	2.6	9.8	1.34
Saturn	120000	1430	95	-190	18	29	1.2	10.2	0.7
Uranus	50000	2800	15	-220	15	84	1.1	10.8	1.58
Neptune	49000	4500	17	-230	2	165	1.4	15.8	2.3
Earth	12700	150	1	20	1	1	1	23.9	5.51
Mars	6800	228	0.1	-20	2	1.88	0.4	24.6	3.95
Pluto	2400	5900	0.003	-240	1	248	-	153.6	2
Mercury	4900	58	0.05	350	0	0.24	0.4	1416	5.4
Venus	12100	108	0.8	470	0	0.62	0.1	5832	5.25

IT tools

Section

2

Looking at the weather

What this is about

In this activity you use a computer database to look for patterns in the weather.

Computers can collect measurements about the weather over a long period of time. They never complain either.

What weather stations collect

If you had a weather station it would collect data using **sensors**. It would automatically take readings from each sensor and store it in memory. The sensors might include a few of these:

A light sensor - to measure the amount of sunlight or show the times of dawn and dusk.

An anemometer - to measure the wind speed.

A wind vane - like the one on a church spire, to see which way the wind blows.

A rain detector - a container that fills and empties when it rains and can measure rain fall.

A humidity sensor - measures the moisture in the air.

A temperature sensor - shows how cold or hot it is.

A pressure sensor - measures changes in the air pressure.

You will need

A database containing weather information collected over a period of time.

A data handling program - perhaps your usual database program or the program that comes with your weather station.

If you have also been collecting weather information manually, you may want to refer to it.

What you can do with your weather data

You can transfer your weather data into a data handling program. These programs allow you to look at the data - to see how things are changing and see which changes seem to occur together.

How has the time of sunrise changed recently? Is it getting earlier or later?

How has the time of sunset changed recently? Is it getting earlier or later?

How has the day length changed recently? Is it getting longer or shorter?

What happens to the length of the day during the month of December (or June)? Why might this be?

Warmth and sun

Which days had the most sunlight?

Which days were the warmest?

Were the warmest days also the sunniest days?

Warmth and rain

Which days had the most rain?

Which days were the coldest?

Were the rainy days the coldest days?

Does it get cold before it rains?

Warmth, rain and wind

Does the wind usually blow from the same direction? Does it blow from several directions?

Which days had the fastest wind?

Were the windy days the wettest days?

Were the windy days the coldest days?

Does the wind blow in a certain direction when it rains?

Pressure, wind and rain

Does the pressure change before it rains?

Is the air pressure high or low when it rains?

Does the pressure change before it gets windy?

Section
2

The chemical elements

What this is about

You use a database of the elements. You then sort and search the information and draw graphs.

A database of the chemical elements stores many facts and figures about the elements. By using a database program you can search for facts and find important patterns in the data.

The order of the periodic table

Use the Sort command to list the elements in order of their:

· Symbols

· Atomic Mass

· Atomic Number

Which of your three lists matches the order in the Periodic table?

Putting elements in their places

Search to find answers to the following:

1. Which elements were discovered in the last 100 years? Where, in the periodic table, are these most often found?

2. Which elements were discovered more than 600 years ago? Where, in the periodic table, are these most often found?

3. Which elements boil at less than 20C? Where, in the periodic table, are these found?

4. Which elements melt at more than 200C. Where, in the periodic table, are these found?

5. Which elements have a density of greater than 4? Where, in the periodic table, are these found?

6. Which elements have a density of less than 1? Where, in the periodic table, are these found?

Finding patterns

Use a scattergraph to look for the following patterns:

1. Is there a periodic pattern between atomic number and melting point?

2. Is there a periodic pattern between atomic number and boiling point?

3. Is there a periodic pattern between atomic number and conductivity?

4. Is there a periodic pattern between atomic number and density?

5. Is there a periodic pattern between atomic mass and density?

Extra patterns

1. Find all the elements which are liquids at room temperature. *Search for elements with boiling points greater than 20C and melting points less than 20C.*

2. Find all the elements in period 1 of the table. How do their melting points change as you go across the periodic table? *Search for the period 1 elements, then plot a graph of melting point against atomic number.*

3. How do the melting points change as you go across period 2? *Search for the period 2 elements, then plot a graph of melting point against atomic number.*

4. How do the melting points change as you go down group one? *Search for the group one elements, then plot a graph of melting point against atomic number.*

5. How do the melting points change as you go down group seven? Is this the same pattern as group one? *Search for the group seven elements, then plot a graph of melting point against atomic number.*

6. What do elements with a cubic body centred crystal form also have in common? Do any of the chemical families have this crystal form exclusively? *Search for cubic body centred crystal forms then look for these on the periodic table.*

IT tools

Section

2

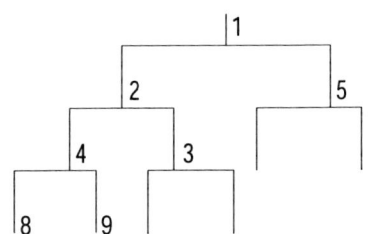

IT tools

Section

2

*T*he branching database is a special kind of database. We use it to classify things and build up a key. It helps us sort out sets of animals, plants and almost anything else. Like ordinary database programs, the branching database stores data. Unlike them, the data is arranged as a branching list.*

The real value of using a branching database comes from getting the pupils to build up a key for themselves. The activity is unlike most others in this book and the reason is to do with why we should do it. We use a branching database not because it helps us produce some end-product or that the product is better or even that it saves time. We use it because it provides a surprisingly strong focus for getting pupils to think about the things we ask them to classify. Our pupils will need to draw on and develop a variety of science skills. They will need to observe, question, discuss and classify. Branching database activities can be good science and even a cornerstone of the best.

A branching database can be made with many of the things that we like to sort out. We might start with musical instruments, animals, sugars, leaves, liquids and materials. We can go on to classify birds, insects, elements, seeds, planets, organs and even bacteria.

The key that the pupils produce is structured like the example below. Despite the easiness of the example, pupils - lower school or upper school - can all use this program to good effect.

1. Does it live in water?
    If yes go to 5   If no go to 2
2. Does it live in the jungle?
    If yes go to 4   If no go to 3
3. ...
4. Does it have a long neck?
    If yes go to 9   If no go to 8
5. ...
8. Is it a lion?
9. Is it a giraffe?
10. ...

Projects with a branching database can be improved with a little structuring. The next few pages show how to do this. There are worksheets for classifying elements and organs of the body. For most abilities, these will be adequate. With younger pupils, say from age 11-12, you might use the more lengthy approach exemplified in Making a branching database - animals **on the next page.**

How to make a branching database

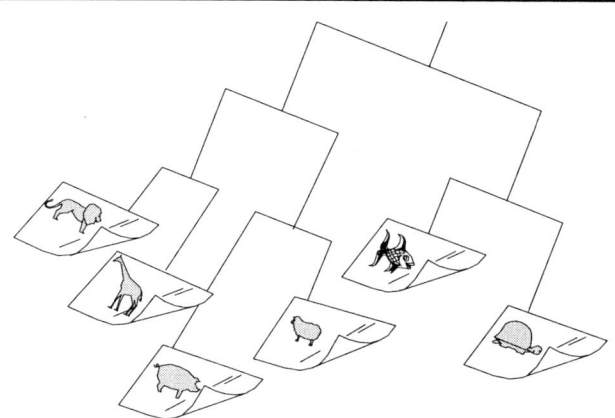

Making a branching database - animals

Play '20 questions' to identify some animals.

Before starting a branching database activity, practise the idea of asking questions which have a yes or no answer: copy pictures of animals from books and magazines and stick them onto cards. Write their names and some key facts on the back of each card. Then play the '20 Questions' game using the cards as stimuli.

Introduce the computer program.

Use a ready-made branching database. Pick up a card, use the program and get the computer to identify the animal. Let the pupils practise this.

Leave the computer.

Get the pupils to sort all the animal cards into sets. Record the reasons for putting things into the sets by labelling them with say, 'Lives in water' on a plain card.

Arrange the animal cards to make a key.

If you wish, use ribbon or paper strips to link the sets together as in the diagram above.

Return to the computer.

Run this part yourself to get off to a healthy start: make a fresh database with two animals, one from two very different sets. Get the group to think of a question to distinguish one animal from the other. A question such as, 'Does it live in water' works better than 'Does it meow'?

Build up the database.

The pupils now pick up one card at a time and work through the program. As they do so the database will grow. The card labels on each set can be to hand as prompts. They should work in groups of two or three. In larger groups things can get boring quite quickly. A group of three can take turns to pick an animal, think of a question and use the keyboard. Pupils should save their work frequently. If they make a mistake it's often easier to get the last good copy back off the disc rather than try to puzzle out what has gone wrong.

Get the others to test the database.

Pupils can try out each others work and suggest ways to improve the questions. You can use the finished database both as a handy key and to introduce the activity next time around.

About the work sheets

Elements - page 30

Pupils first play a twenty questions 'game' about the elements. They then use a branching database program to build up an element identification key.

The activity would be suitable for any group of pupils who had a reasonable knowledge of 10 to 20 elements. The pupils might appreciate using a set of 'cards' with the element names on. For less knowledgeable groups, you can write some properties of the elements on the cards.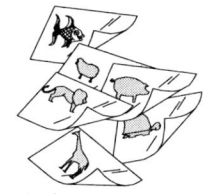

To get off to a reliable start, 'prime' the computer with two elements before the pupils begin. That is, get the Sorting game program running and start a new game called **Elements**. You might then enter the two elements chlorine and sodium and the question "is it a gas?"
IT level: easy

Organs - page 31

Pupils first play a twenty questions 'game' about the parts of the body. They then use a branching database program to build up an identification key.

The activity involves careful thought and observation. It would be suitable for almost any group of pupils. The pupils might appreciate having a model of the body nearby. Advanced groups can be told to focus on organ function.

As the instructions show, it is quite important not to enter the wrong questions and answers. If a mistake is entered, try to avoid saving it on disc.

To get off to a reliable start, 'prime' the computer with two organs before the pupils begin. Simply, get the Sorting game program running and start a new game called **Organs**. You might then enter two organs, the stomach and the heart and the question "is it part of the digestive system?"
IT level: easy

IT tools

Section
2

Classifying the elements

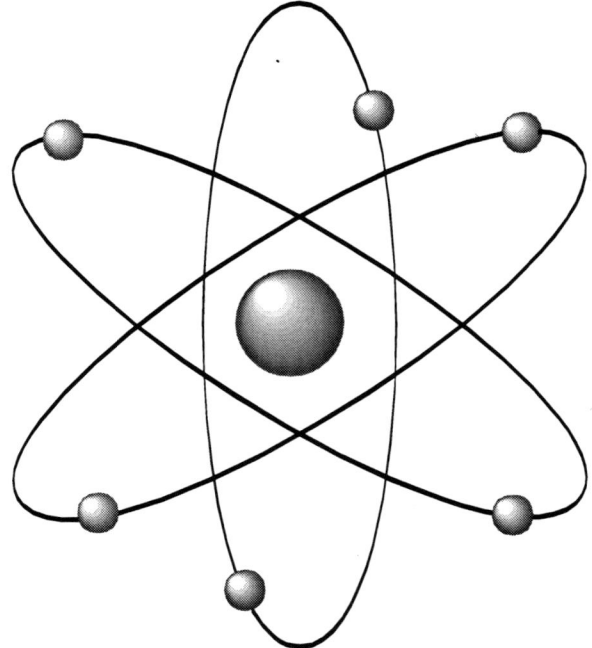

What this is about

You use a branching database program to classify the elements and build an element identification key.

Elements are the building blocks for all the things you will ever meet. Most elements have properties that help you to identify them quite easily.

In this activity you will play a 'game' where you have to guess the identity of an element.

Playing a game

Play the game *Twenty questions*. In this game, one of your group plays the 'thinker'. The others have to guess the element they are thinking of.

Note to the 'thinker':

You have to think of an element. The periodic table lists a hundred of them, but choose one you have studied in class. The others have to guess what your element is by asking questions. You can only answer their questions with a 'yes' or a 'no'.

Note to the players:

You have to guess the identity of an element by asking the 'thinker' questions. Your questions will only get a yes or no answer. To make the game more scientific, you have to ask questions about how the element looks and how it behaves. For example,

You may ask, "is it a green" but not "does it begin with..."

You may ask, "does it react with water?" but not "is it reactive?"

You may ask, "is it used for ... ?" but only if you are stuck.

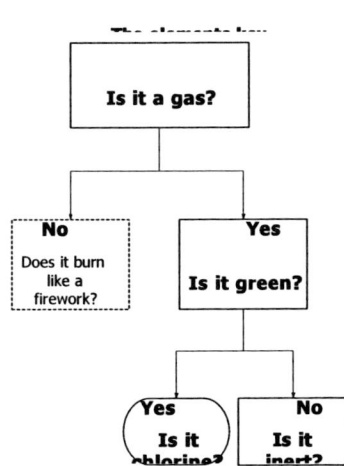

Playing the computer sorting game

Get the *Sorting game** program running.

Use the old game called *Elements* on the disc and start the game.

Think of an element and answer yes or no to the questions.

If the computer guesses your element correctly, think of another element and play the game again.

If the computer gives up and asks you what it is, tell it. Then, very carefully, follow the instructions on the screen.

Save your work on the disc from time to time. Do not save any mistakes on the disc.

Continue until your time is up.

Note

*These are also known as branching database, tree database, sorting game, dichotomous key or binary key programs. For example:

Retreeval (For Acorn from Kudlian Soft, 8 Barrow Road, Kenilworth, Warwickshire, CV8 1EH. Telephone/Fax: 01926 851147) – clever branching database program for classification projects.

Sorting Game (PC/Acorn, from MAPE, Newman College, Bartley Green, Birmingham, B32 3NT) – the original program.

Window Tree (for PC –SITSS, Bourne House, Radbrook, Shrewsbury, SY3 9BJ. Telephone: 0743 246043 Fax: 0743 368481

IT tools

Section
2

Identifying the parts of the body

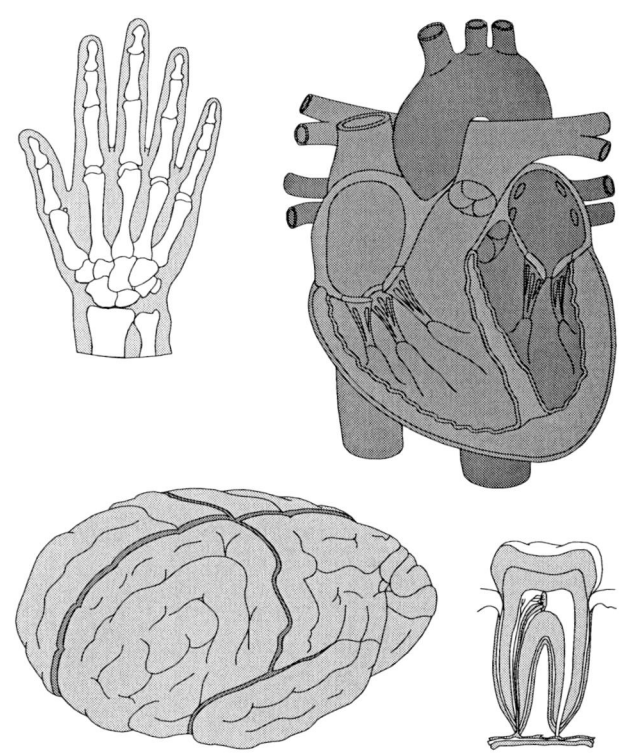

What this is about

You use a branching database program to build an identification key for the parts of the body

Organs are the major units of the body. Heart, lungs, kidneys - even skin is an organ. They have different functions or jobs to do.

In this exercise your knowledge of the organs will be put to the test. You will play a 'game' where you have to guess the identity of a body organ.

Playing a game

Play the game *Twenty questions*. In this game, one of your group plays the 'thinker'. The others have to guess the organ they are thinking of.

Note to the 'thinker':

You have to think of an organ. Choose one you have studied in class as you will need to know something about it. You might choose from heart, liver, lungs, kidneys, skin, intestine, oesophagus, pancreas, stomach, duodenum, rectum, ear, eye, nose or brain. The others have to guess what the organ is by asking questions. You can only answer their questions with a 'yes' or a 'no'.

Note to the other players:

You have to guess the identity of an organ by asking the 'thinker' questions. Your questions will only get a yes or no answer. To make the game more scientific, you have to ask questions about what the organ does, what it looks like or where it is found in the body. For example:

You may ask: "does it digest food" but not "does it begin with 's' "

You may ask: "is it as big as a grapefruit?" but not "is it small?"

You may ask "would you find it in the abdominal cavity ... ?" but not "would you find it in a meat pie?"

Playing the computer sorting game

Get the **Sorting game**[*] program running.

Use the old game called **Organs** on the disc and start the game.

Think of an organ and answer yes or no to the questions. Note:

If the computer guesses the organ correctly, think of another and play the game again.

If the computer gives up and asks you what it is, tell it. Then, very carefully, follow the instructions on the screen.

Save your work on the disc from time to time. Do not save any mistakes on the disc.

Continue until your time is up.

Note

*These are also known as branching database, tree database, sorting game, dichotomous key or binary key programs.

IT tools

Section

2

	A	B	C
1	**Body**	**Gravity**	**Your weight**
2	Earth	1.0	40
3	Moon	0.2	6
4	Mercury	0.4	15
5	Mars	0.4	15
6	Venus	0.9	34
7	Uranus	1.0	40
8	Saturn	1.1	44
9	Neptune	1.5	60
10	Jupiter	2.6	104
11			
12	MAXIMUM	2.6	104
13	MINIMUM	0.2	6.4

IT tools

Section

2

A spreadsheet is very much a multipurpose program. You can use one as a ready-made results table and quickly produce a graph from it. You can also use one as a data handling program sorting and searching your data and again producing graphs from it. But perhaps the most interesting feature of a spreadsheet is its potential for doing calculations and 'mathematical modelling'.

If you had, for example, some information about the gravity force on various celestial bodies you could get the spreadsheet to work out how much you would weigh on each of them. First, you enter the information and then you write formulae which do the maths. There is nothing really special about these formulae - it's merely algebra. For example, in the spreadsheet here, the box or cell C3 calculates your weight on the moon and contains the formula C2 x B3.

This 'sheet' is a simple mathematical model - it is a mathematical alternative to actually going to the moon to weigh yourself. You only need to enter your weight in cell C2 to see it calculated for different celestial bodies.

Spreadsheets have an astonishing range of functions that can help with maths or modelling. They can total or average columns, look for maximum or minimum values and turn any mathematical trick.

You can use a spreadsheet to build models as complicated as you wish. You can model the gas laws, chemical equilibrium and even the Hardy-Weinberg distribution law. It is this, the modelling side of using spreadsheets, that can make them really useful in science.

The following pages illustrate how a spreadsheet can begin to be used in science. The ideas aim to exemplify their role in organising, recording and analysing data - all of which are key features of exploring science.

Spreadsheet glossary

Bar chart - a graph plotted with a bar for each item of data.

Cell reference - every cell has a reference code you can use to refer to the cell. The spreadsheet columns might be labelled from A to Z while the rows might be labelled from 1 to 100. Cell references are used in building spreadsheet formulae - for example, the following formula works out a percentage: *=100 * B2/B3*

Cells - boxes on the screen into which you type.

Copy - a feature which copies something you've typed in on the spreadsheet, saving you having to key it in again.

Data - the numbers and words that you've collected together to store and study.

Database program - like a spreadsheet this a program to handle data. This raises the question of which is best to use for the job of handling some data. Some say that you should use a spreadsheet if you can visualise your data as a grid or table, if your data is mostly numbers, if you want to do calculations or if you want to see all your data in one glance. When there is lots of data to search through, the advice is to use a database program instead.

Formula - you make a formula when you need to do some maths, such as multiply two numbers together. A formula to multiply two numbers together might look like *B2*C2.*

Function - a formula which is built into the spreadsheet. It can work out the total or average of a column.

Goal Seek - a term used to describe using a spreadsheet to ask "what value must this change to, to make this value...?". In other words you can work backwards from a graph to your original data. (We used to call this 'fiddling results'). It can be very useful though you might not ever use the feature in class.

Labels - the headings for the spreadsheet data.

Line graph - is like a bar chart. A dot or cross is plotted for each item of data.

Move - a feature which moves something you've typed in to a new place on the spreadsheet. It saves you having to key it in again.

Name - you can highlight a cell and give it a useful name. You might call one cell 'distance' and the other cell 'time'. You could then type in a formula that made more sense, i.e. to calculate a speed, you would type =distance/time instead of those awkward formulae.

Operator - meaning arithmetic operators such as multiply, divide, add and subtract or * / + - .

Pie Chart - a chart which can show the relative amounts of, say oxygen in the air. The results are shown as a percentage.

Range - is a set or range of cells. In the function *=MAX(C3:C8)*, the range is the set of cells stretching from cell C3 to C8.

Scattergraph - an x-y graph where you can compare two variables to find a pattern.

Sort - a feature which sorts a column into order. You mark or select the column you want to sort. You will be asked if you want to sort it into ascending or descending order. Words are sorted into alphabetical order. Note that the computer 'alphabet' begins with numbers so that in an address: *7 Heathview* comes before *Heathview*. Similarly when numbers get sorted into alphabetical order, not always by mistake - the result would be that 11 comes before 7.

Values - this is your data i.e. the numbers and words that you've collected and entered on the sheet.

Spreadsheet Jargon

	A	B	C
1	▾		
2		Gravity %	Your weight
3	Earth	100	40
4	Moon	20	6
5	Mercury	40	15
6	Mars	40	15
7	Venus	90	34
8	▾ Uranus	100	40
9			
10	MAXIMUM	100	40
11	MINIMUM	20	6

Cell Reference Each cell has a **cell reference**. This is A1

Labels Headings for the table.

Cell Each box in the table is called a **cell**.

Values The data you've collected and typed in.

Formula This cell does some maths using a **formula**.. Your weight gets worked out with the formula: B5*C3

Function a built-in formula. This cell will find the maximum weight. This function looks like: MAX(C3:C8) You can also do averages.

IT tools

Section
2

Spreadsheets - teaching notes

Measuring the energy in food - page 36

Pupils have to plan their experiment and then enter their results in the spreadsheet. This calculation always seems to cause unnecessary confusion and it would be in order to prepare the spreadsheet for the pupils.
Skills used: record, calculate and draw a bar chart. IT level: Medium

Nutrition and breakfast cereals - page 37

The pupils do a survey of breakfast cereals and use the spreadsheet to handle the data. This can be successful with a survey of as few as five cereals, but a bigger sample is preferable. (We found that the fibre content is inversely related to the energy content - but seriously wondered if this was significant.) If your survey is extensive you may find that a database program does the job more easily. Double check the data as mistakes are very easy with this amount of data. You might also record the size and price of the packet to work out how much energy or protein you get for your money.
Skills used: record, calculate and scattergraph. IT level: Medium

A population of wolves and deer - page 38

An introduction to drawing a scattergraph. To save a class time you can prepare the data in advance and leave the pupils to draw the graph.
Skills used: record and draw a scattergraph. IT level: Easy

Soil water and organic matter - page 39

This exercise uses a spreadsheet to do some fairly easy calculations on the results of a soil analysis. Easy they may be, they are still a puzzle. You can use the spreadsheet to average the class' results. Or if you tested different soil samples you could use the spreadsheet to draw a bar chart to compare them.
Skills used: record and calculate. IT level: Medium

How our use of fuels has changed - page 40

An introduction to drawing a scattergraph. To save a class time you can prepare the data in advance and leave the pupils to draw the graph.
Skills used: record and draw a scattergraph. IT level: Easy

Forces: testing cotton reel rollers - page 41

This is a fairly simple spreadsheet activity. Pupils use the spreadsheet to record their results. The hardest part is interpreting the scattergraph.
Skills used: record and draw a scattergraph. IT level: Easy

Forces: testing bridge designs - page 42

This is a very simple spreadsheet activity suitable for first timers.
Skills used: record, sort and draw a bar chart. IT level: Easy.

Distance, time and speed - page 43

This spreadsheet on calculating speed from time and distance values was used successfully with a fairly able year 10 group with average IT skills. Even so, in some cases you will want to prepare the spreadsheet in advance.
Skills used: record, calculate and draw a scattergraph from two discontinuous series. IT level: Medium

Boyle's law - page 44

This is a fairly simple spreadsheet activity consisting of entering the pressure and volume readings from an experiment.
Skills used: record, calculate and draw a scattergraph from two discontinuous series. IT level: Easy/Medium.

Energy: home insulation I - page 45

Uses a spreadsheet to work out how much home insulation can save. The first part of the exercise simply involves collecting information from the page to enter in the spreadsheet, nevertheless, there is rather too much here for novices.
Skills used: record, calculate and draw a standard or stacked bar chart. IT level: Medium

Energy: home insulation II - page 46

This spreadsheet uses the finished work from the previous exercise. As above, the maths is relatively easy, but do consider going through this as a paper exercise first. The extra section ends with an interesting project - pupils will need to work out if living in the basement or top flat would be the most economical. They will need to make up a separate spreadsheet for each flat and compare them.
Skills used: record, calculate and draw a bar chart. IT level: Medium/ Hard

IT tools

Section

2

Gravity in outer space - page 47

This is an introduction to using a spreadsheet. This is a fairly self-contained activity. If the pupils have good word processing skills, this would be a good opportunity to introduce the idea of cutting and pasting data from the spreadsheet to the word processor.
Skills used: calculate, sort and draw a bar chart. IT level: Easy

Force, mass, acceleration: cars - page 48

This sheet provides several starting points for exploring data on cars. The pupils need to plot scattergraphs in the search for patterns and convert units to SI. They might even try to see if the relationship F=ma holds true. They will need to create new columns with formulae of their own devising. They should also take into account and discuss the over-simplification in this model.
Skills used: record, calculate and draw a scattergraph from two discontinuous series. IT level: Hard

Analysing data I - page 21

This is a data handling exercise on analysing pupils' heights, weights and so on. It looks at the choice of graphs available and if pupils have some data they can try the ideas on the sheet. This page can be found in the previous section on **Databases**.
Skills used: draw pie and bar graphs. IT level: Easy/Medium

Analysing data II - page 22

This too is a data handling exercise on analysing pupils' heights, weights and so on. It looks at averages, sorting and drawing bar graphs and scattergraphs. With these tools they can spot trends and patterns. This page can be found in the previous section on **Databases**.
Skills used: average, sort, draw bar graphs and scattergraphs. IT level: Medium

Teaching progression points

The spreadsheet is probably the most versatile data handling tool we have. We can take the many skills involved in using one and put them into a 'progression ladder' as shown below.

Using modern software a pupil will work their way through the first half of this table very quickly indeed.

After that things get harder. The skills at the bottom of the table are very hard to develop with a spreadsheet. It might even be better not to use a spreadsheet at all here. Perhaps more accessible tools, **model builders** and **science programs** could be used instead.

Nevertheless, pupils will benefit from plenty of exposure to spreadsheets. They still remain very valuable for so many routine tasks in science.

Progression in spreadsheet skills
Use a spreadsheet as a recording table. Enter data.
Sort a table into order. Prepare pie or bar graphs
Use a spreadsheet with hidden (i.e. prepared) formulae.
Graph one variable against another.
Change spreadsheet formulae to see how the model is affected.
Design and use a computer model
Appreciate the need for computer models
Use a computer model and derive the relationship between its variables
Evaluate a model: the need for it, its effectiveness and appropriateness. Suggest improvements.
Design, implement, justify use of a computer model

IT tools

Section

2

Measuring the energy in food

What this is about

Using a spreadsheet to calculate the energy produced by different foods.

Different foods contain different amounts of energy. An oily peanut is full of energy while a lettuce leaf has hardly any. This energy, chemical energy, is in the chemicals that make up food. If we burn the food, its chemical energy is converted into heat energy. If we use this energy to heat water and measure its rise in temperature we can find how much energy the food contains.

Plan

Plan an investigation to measure the chemical energy content in different foods. The diagram provides you with some clues.

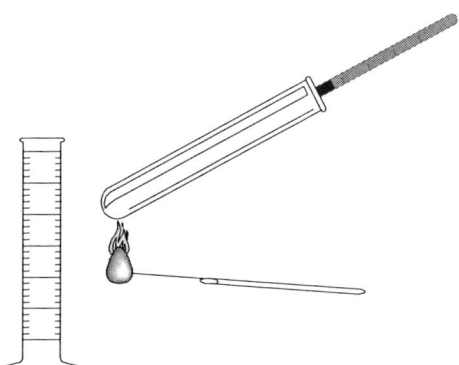

Before you start you should ask yourself:

- Should you weigh your food before burning?
- Should you weigh your food after burning?
- Will any of the food energy be wasted when you burn the food? How can you avoid this?
- How much water will you heat?
- What else will you measure?
- How can you and the class get to test several different foods during the lesson?

What to do

Do your experiment.

Open the spreadsheet **Food** or follow the instructions in the diagram to make your own spreadsheet. This should take you about 10 minutes.

Record your results in columns **D**, **E** and **F**.

Draw a graph to show which foods have the most and least energy. See the diagram for details of how to draw a graph.

What does the graph tell you?

Questions

List the foods you tested in order of their energy content.

Compare your results with values in a Calorie table.

Do your results compare well with the Calorie table? What are the causes of error in your experiment?

Get another book and compare the two sets of printed results. Do these results compare well or is there any reason why the books might not show the same values?

	A	B	C	D	E	F	G	H	I	J
1		Unit	Example	Peanut	Crisp	Pea				
2	Mass of food	g	1.6							
3	Mass of food after burning	g	0.5							
4	**Mass of food burned**	g	**1.1**							
5	Amount of water heated	cm3	10							
6	Starting temperature of water	deg C	20							
7	Final temperature of water	deg C	31							
8	**Temperature rise**	deg C	**11**							
9										
10	**Energy content of food**	kJ/100g	**508.2**							
11										
12										
13										
14										
15										
16										
17										
18										
19										
20										

How to set up your spreadsheet
1. Copy row 1 as shown here.
2. Copy columns A and B also as shown here.
3. Move to cell C4 and enter the formula =C2-C3
This works out the mass of food burned.
4. Move to cell C8 and enter the formula =C7-C6
This works out the temperature rise.
5. Move to cell C10 and enter the formula =4.2*C4*C8*100/C5
This works out the energy content of the food.
6. Copy cells C4 to C10. Paste them into cells D4 to F4.
7. Enter your results in columns D, E and F.

How to draw a graph with your spreadsheet
1. Highlight cells D1 to F1
2. Hold down the CTRL or 'special' key.
3. Highlight cells D10 to F10
4. Get the program to draw a bar graph.

IT tools

Section 2

Nutrition and breakfast cereals

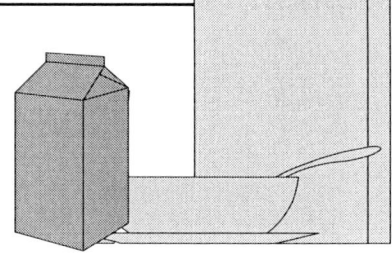

What this is about

Using a spreadsheet to record and calculate the nutritional content of different foods.

Breakfast cereals have a wide range of ingredients. You probably choose yours because you like it - but what would you do if you wanted to improve your diet? How would you set about comparing breakfast cereals?

In this activity you will do a survey of some cereals. You will use a computer spreadsheet to help you look at the results.

What to do

Collect your information about cereals. Make a note of the energy, protein, carbohydrate, sugars, fat, fibre and sodium per 100g of cereal.

Find the **Cereals** spreadsheet on your disc - otherwise, you can make your own using the diagram below.

Enter the information you collected into your spreadsheet.

Quick questions

1. Which cereal has the highest sugar content? Does this also have the highest energy content?

2. Which cereal has the lowest sugar content? Does this also have the lowest energy content?

3. Which cereal has the highest carbohydrate content? Does this also have the highest energy content?

4. Which cereal has the lowest carbohydrate content? Does this also have the lowest energy content?

5. Have you found any patterns in your data?

Patterns

You might ask,

Do cereals with the most sugar have the most energy?

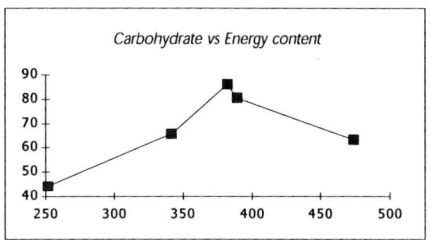

One way to answer this is to draw a graph:

Highlight the **Energy** column. Then hold down CTRL and highlight the **Sugar** column. Use the Chart command to plot a scattergraph. *What does your graph tell you?*

Plot more graphs to answer these questions:

Do cereals with the most carbohydrate have the most energy?

Do cereals with the most fat have the most energy?

Do cereals with the most fibre have the least energy?

Carbohydrate vs Energy content

Total nutrient

The total nutrient of the cereals is the sum of their carbohydrate, protein, fat, fibre and sodium. You can calculate this as shown in the diagram below.

Your totals do not add up to 100g. Suggest what the rest of the cereal might be made of.

Note

You need to find out how your spreadsheet draws a scattergraph from columns that are not side-by-side. Normally you highlight the first column, then hold down the CTRL or 'special' key to highlight the second column.

	A	B	C	D	E	F	G	H	I	J	K
1	Cereal	Energy kJ	Total Carbohydrate	Protein	Fat	Fibre	Sodium	Sugar	Total		
2	Bran worms	252	44	13	3.5	28	3.4	15.5	**91.9**		
3	Weetabix	342	65.8	11.2	2.7	12.9	0.3	4.9	**92.9**		
4	Puffed wheat	382	86	6.3	1.4	0.8	1.2	10	**95.7**		
5	Weetos	389	80.5	6.2	4.5	4.6	0.3	35.5	**96.1**		
6	Crispy clusters	474	63.3	8.7	20.5	3.6	0.3	22.1	**96.4**		
7											
8											
9											
10											
11											
12											
13											
14											
15											
16											

Breakfast cereal survey - figures as g per 100g of cereal
Tip: Keep the cereal packets handy to check your typing.

How to set up your spreadsheet
1. Enter the headings in row 1
2. Enter your cereal names in column A
3. Enter your cereal data in the rest of the table.
How to do calculate the total nutrient content
Move to cell I2.
Type =SUM(C2:G2)
Copy cell I2 Paste it into cells I3 to I6.

IT tools

Section

2

A population of wolves and deer

What this is about

Using a spreadsheet to prepare and label graphs of some data on a population of animals.

In a North American park lived a population of deer. They fed on a good supply of grass and tree bark - in fact, there was enough food to sustain 1500 deer.

The park rangers always kept a careful eye on them - keeping a record of how many there were, how many died and so on. In the same park, wolves roamed and preyed on the deer. One year the park rangers decided to kill off a pack of wolves...

In the spreadsheet below, you will find the rangers' records. Your job is to use the records and come to some conclusions about the management of the park.

What to do

Look at the year 1970 in the table of data:

	A	B	C	D	E	F
1	**Wolves and deer**					
2		Deer alive at the start of the year	Deer died of sickness or old age	Deer killed by wolves	Deer born this year	Deer alive at the end of year
3	1970	1000	100	100	205	1005
4	1971	1005	95	110	215	1015
5	1972	1015	110	105	200	1000
6	1973	1000	110	115	205	990
7	1974	990	95	105	215	1005
8	1975	1005	105	50	250	1100
9	1976	1100	110	10	220	1200
10	1977	1200	120	0	240	1320
11	1978	1320	120	0	250	1450
12	1979	1450	110	0	290	1630
13	1980	1630	150	0	320	1800
14	1981	1800	1000	0	100	900
15	1982	900	300	0	50	650
16	1983	650	150	0	50	550
17	1985	550	80	0	70	540

How many deer were alive at the beginning of the year?

How many deer were alive at the end of the year?

How many deer were preyed on?

How many deer died of natural causes?

How many deer were born that year?

Analysing the data

Open the spreadsheet file **Deer** containing the data below. If you cannot find the file, open a new spreadsheet and enter the data. This should take you about 10 minutes.

Plot a scattergraph using the data. To do this you highlight all the data and choose the chart command. Make sure that your scattergraph:

Has a key or legend to label the lines.
Has a labelled vertical axis for the number of deer, and a labelled horizontal axis for the year.
Fits on the screen but is much bigger than this:

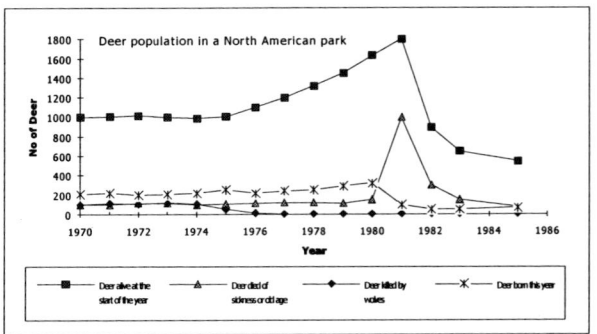

Questions

One year the park rangers decided to kill a pack of wolves. Look at your graph and answer:

In which year do you think this was?

What happened to the total deer population between 1980 and 1981?

Why do you think so many deer died in this year?

What was happening to the number of young deer being born?

In 1985, the park rangers decided to do two things. They planted more trees and grass in the park AND they allowed some wolves to live in the park in future. Why did they decide to do these things?

Do you agree with their course of action?

Note
Source could not be traced.

Analysing soil for water and organic matter

What this is about

Using a spreadsheet to calculate the amount of water and organic material in soil.

Soil might be a nuisance when it gets on your clothes but to plants it's vital for life.

Soil contains a liquid part which is, of course, water. Can you think of a way of measuring how much of soil is water?

Soil also contains a solid part with organic and inorganic material. The organic matter is easily turned into gas by fierce heat or burning. Can you think of a way of measuring how much of soil is organic matter?

What to do

Do an investigation to measure the water content of soil.

Also measure how much of the soil is organic matter.

Open the spreadsheet file **Soil**. Otherwise, follow the instructions in the diagram to make your own spreadsheet. This should take you about 10 minutes.

Enter your results in the spreadsheet, but take care not to type in the shaded areas you can see below. These areas contain the formulae for some calculations.

You can also collect and enter the results from other groups in the class.

Questions

1. What happened to the soil water when you did this experiment?

2. What sort of soil would contain lots of water? What sort of soil would have little water?

3. What happened to the organic matter when you heated the dry soil?

4. What sort of soil would contain lots of organic matter?

5. What sort of soil would have little organic matter?

6. How do the class' results compare?

Finding the right answer

When several groups measure the same soil sample, we use the average result. *Use the right-most column of your spreadsheet to average the class' results:*

In cell G8 enter the formula =AVERAGE(B8:F8)
In cell G14 enter the formula =AVERAGE(B15:F15)

Is this a more correct answer? Are any results far from the average result?

Note

If you were comparing different soil samples you could use the same spreadsheet to draw a bar graph to compare them. To do this highlight cells B1 to F1. Hold down the CTRL key and highlight cells B8 to F8. Get the program to draw a bar chart.

	A	B	C	D	E	F	G	H	I
1	Water content	Result 1	Result 2	Result 3	Result 4	Result 5			
2	Mass of dish g	250							
3	Mass of dish and wet soil g	355							
4	Mass of wet soil sample g	105							
5	Mass of dish and dry soil g	301							
6	Mass of dry soil sample g	51							
7	Mass of water g	54							
8	Percentage water	51							
9									
10	Organic matter								
11	Mass of dish and dry soil g	301							
12	Mass of dish and burnt soil sample g	275							
13	Mass of inorganic matter remaining g	26							
14	Mass of organic matter g	25							
15	Percentage of organic matter	49							
16									
17									

How to set up your spreadsheet

1. Copy row 1 as shown here.
2. Copy column A also as shown here.
3. Move to cell B4 and enter the formula =B3-B2
   This works out the mass of the wet soil sample.
4. Move to cell B6 and enter the formula =B5-B2
   This works out the mass of the dry soil sample.
5. Move to cell B7 and enter the formula =B4-B6
   This works out mass of water in the soil.
6. Move to cell B8 and enter the formula =100*B7/B4
7. Move to cell B11 and enter the formula =B5.
8. Move to cell B13 and enter the formula =B11-B12
9. Move to cell B14 and enter the formula =B6-B13
10. Move to cell B15 and enter the formula =100*B14/B6
11. Highlight cells B2 to B15 and choose Copy.
12. Highlight cells C2 to F2 and choose Paste.

IT tools

Section 2

How our use of fuels has changed

What this is about

Using a spreadsheet to prepare and label a graph of some data on fuels.

We use fuels to get the energy we need for heat, electricity and transport. These fuels include coal, oil and gas. Fuels such as oil are also used by the chemical industry. Instead of burning it they turn it into plastics and chemicals. In the last 20 years our use of fuels has changed. The table below shows this:

	A	B	C	D	E	F
1	How has our use of fuels changed?					
2	Fuel	1960	1965	1970	1975	1980
3	Town Gas	4	4	3.8	0.5	0
4	Coal	24	18.5	13	7	5
5	Natural gas	0	0	1	12	16
6	Petrol/Oil	14	20	26	24	23
7	Totals					

What to do

Find this data table on your disc - otherwise enter the data into a spreadsheet program.

Highlight all the data from **cells A2 to F7**. Then get the spreadsheet to draw a scattergraph of the data:

Make sure your graph has labels on the horizontal and vertical axes.

Add a legend (or key) to show which graph is which fuel - unlike our graph below.

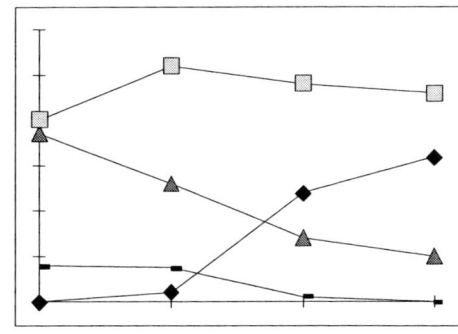

Questions

Which fuels were available to us 30 years ago?

How has this changed since then?

Which fuel has become increasingly popular?

Which fuels have become decreasingly popular?

Town gas or 'gas' was made from coal. What has steadily replaced it?

Extra

How has our energy use changed over the years? Can you suggest any reasons for this? *Use your spreadsheet to work out the total amount of energy used in each year:*

In cell B7 enter the formula =SUM(B3:B6) Copy cell B7. Paste it into cells C7 to F7

In the early seventies the world price of oil soared. How is this event shown on your graph? *Plot a graph of total energy against the year:*

Highlight cells B2 to F2. Hold down CTRL key while you highlight cells B7 to F7 Get the program to draw a scattergraph

Notes

Source of data: Middle School Science Resources, Energy Unit, Heinemann / ILEA.

Forces: testing cotton reel rollers

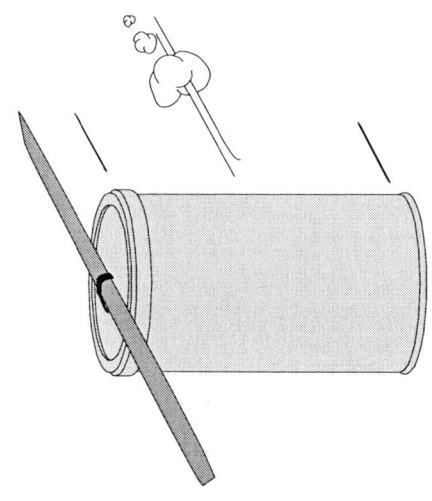

What this is about

Using a spreadsheet to average your results and produce a bar chart.

Have you ever made one of those cotton-reel rollers powered by elastic bands? When you wind them up you store energy in the elastic band. When you let them go this energy is released and the roller moves forward. In this activity you will be taking a more scientific look at this toy.

For example, do you think that if you turned the band twice as far, the roller will go twice as far? Is there a pattern between the number of times you turn the band and how far it travels.

You will need to investigate to find out.

What to do

· Make a cotton reel roller.

· Wind up your roller different numbers of turns and measure how far it travels. Do this three times for each number of turns.

· Remember to make your investigation fair. Remember to be fair too: let your partner have a go!

Your results

Make a spreadsheet and enter your results as shown below. The spreadsheet will calculate the average of your three 'goes' with the roller.

Plot a scattergraph of your results as shown here:

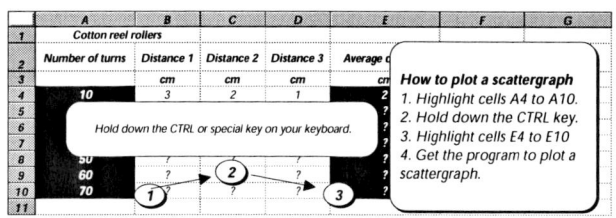

How to plot a scattergraph
1. Highlight cells A4 to A10.
2. Hold down the CTRL key.
3. Highlight cells E4 to E10
4. Get the program to plot a scattergraph.

Is there any pattern in the results? In other words, does increasing the number of turns always increase the distance travelled?

Write a report of your investigation for next year's class. Explain how they should do it.

If you had tested an elastic band powered 'plane, would you expect to find a similar pattern?

	A	B	C	D	E	F	G
1	Cotton reel rollers						
2	Number of turns	Distance 1	Distance 2	Distance 3	Average distance		
3		cm	cm	cm	cm		
4	10	3	2	1	2		
5	20						
6	etc						
7	etc						
8	etc						
9	etc						
10	etc						
11							
12							
13							
14							
15							

How to set up your spreadsheet
1. Enter the headings in rows 1, 2 and 3.
2. Enter the number of turns of the elastic band in column A.
3. Measure how far the roller goes and record your reading in column B.
4. Do this for different numbers of turns.
5. Repeat this twice more. Enter your readings in column C and D.
6. Move to cell E4 and enter the formula =AVERAGE (B4:D4). This works out the average of your three results.
7. Copy cell E4. Paste it into cells E5 to E10.

IT tools

Section
2

Forces: testing bridge designs

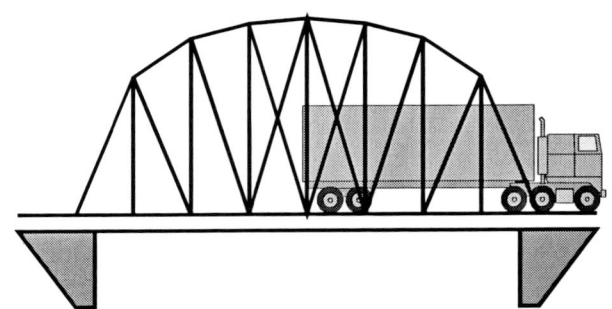

What this is about

An introduction to using a spreadsheet to predict results.

Bridges have to carry huge loads - cars, lorries and more. Now and again you hear about a bridge disaster, so we don't always get things right.

Maybe we should make our bridges stronger? If we used more material the bridge would be stronger, but could we predict how strong a bridge will be without actually making it? In this activity you build some bridges and test them with weights. Then you can advise on how strong they should be made.

You will need

Tape, scissors, weights, 'piers' and soft card.

What to do

1. Use one piece of card to build a bridge to span a 15 cm gap. Here are some bridge design ideas:

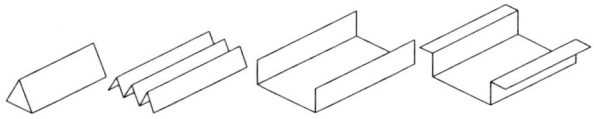

2. Test your bridge with weights to see how much it can take.

3. You could make the same bridge using two, three, four and five pieces of card. First, predict and write down how much weight each of these bridges might take.

4. Make these bridges and test them with weights. Make sure you test your bridges scientifically.

Results

Record your predictions and results in a table. *Open a spreadsheet as shown in our example:*

How does using more material affect the strength of a bridge? *Highlight your results and draw a bar chart of your results. For example, it might look like this:*

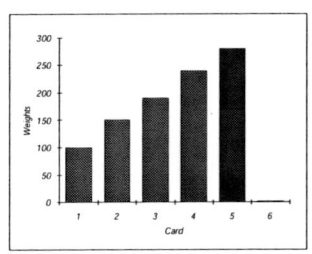

How many weights would a bridge with six or seven pieces of card take?

Get your graph to help you find the answer. Draw a line through the tops of your bars in the chart. Try to read the answers from the graph. For example,

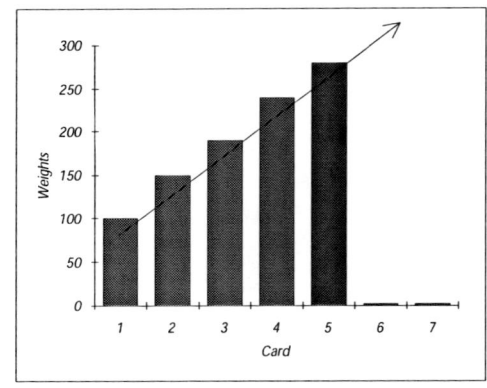

Write a note to a bridge engineer. Describe your thoughts about using more material to build a bridge.

	A	B	C
1	**Number of**	**How many weights**	**Actual number**
2	sheets of paper	might it take?	of weights taken
3	1	-	
4	2		
5	3		
6	4		
7	5		
8	6		Do not make
9	7		Do not make

IT tools

Section

2

Distance, time and speed

What this is about

Using a spreadsheet to do some calculations and draw a graph.

It was in Rome, 1987 that Ben Johnson smashed a world record and dashed to fame. He ran 100 metres in just 9.83 seconds. His times, throughout the race, are shown here. Just how fast was he running? And how did his speed change during the race?

You can work out Ben's average speed quite easily.

$$Speed = \frac{Distance}{Time\ taken} = \frac{100}{9.83} = 10.17\ m/s \quad (About\ 22\ mph)$$

That figure is Ben's average speed for the whole race. But did he start the race slowly and then get faster? And did he ever slow down during the race? The way to find out is to work out how fast was he running at each step of the race.

You may be able to work this out for yourself. On the other hand, a spreadsheet can do this and draw a graph for you too.

Time taken s	Distance run m
0.00	0
1.65	10
2.76	20
3.71	30
4.63	40
5.52	50
6.38	60
7.23	70
8.09	80
8.96	90
9.83	100

Calculating speed

Calculate how fast Ben was running at each step of the race. *Follow the instructions in the panel below.*

How fast was Ben travelling at the 50 metre mark? *Look at the speeds in the spreadsheet to find out.*

What was his fastest speed during the race?

At what point in the race was he running the fastest?

Seeing how the speed changes

Describe how Ben's speed changed during the race. *Follow the instructions in the panel on how to draw a graph.*

Is there any room for improvement in Ben's pattern of speed?

Note

As the graph is plotted from a discontinuous set of cells, hold down the CTRL (or special) key to highlight the second column of cells.
Source: Science Scene, Teacher's Resource Pack 3 (Hodder & Stoughton)

	A	B	C	D	E	F	G	H	I
1	Time taken s	Distance run m	Distance travelled m	Time taken for that part of the race	Speed for that portion of the race				
2	0.00	0	0	0.00	0.0				
3	1.65	10	10	1.65	6.1				
4	2.76	20							
5	3.71	30							
6	4.63	40							
7	5.52	50							
8	6.38	60							
9	7.23	70							
10	8.09	80							
11	8.96	90							
12	9.83	100							
13									
14									
15									
16									
17									
18									

How to set up your spreadsheet
1. Copy row 1 and 2 as shown here.
2. Fill columns A and B (see the results table)
3. Move to cell C3 and enter the formula =B3-B2
This works out the distance travelled.
4. Move to cell D3 and enter the formula =A3-A2
This works out the time taken for this part of the race.
5. Move to cell E3 and enter the formula =C3/D3
This works out the average speed.
6. Copy cells C3 to E3. Paste them into cells C4 to C12.

How to draw a graph with your spreadsheet
1. Highlight cells B2 to B12.
2. Hold down the CTRL or similar key.
3. Highlight cells E2 to E12
4. Get the program to draw a scattergraph.

IT tools

Section

2

Boyle's law

What this is about

Using a spreadsheet to draw a scattergraph and 'model' or calculate the results of an experiment.

A scientist called Boyle tried to make sense of the fact that the more we squash a gas, the more it seems to fight back.
He did an experiment where he squashed a gas and measured its volume and at the same time measured how much pressure it developed. He seemed sure that there must be a pattern in the measurements he collected.
You might try a similar experiment. Unlike Boyle, you have a computer to help you deal with all those numbers.

Do your experiment

You will need the results of an experiment like Boyle's. For example, here are Linda Webb's results:

Pressure x100,000N/m2	Volume cm3
3.13	16.0
2.97	17.0
2.78	18.2
2.56	19.8
2.42	21.0
2.21	23.0
1.97	26.0
1.83	28.0
1.60	31.7
1.40	36.0
1.20	41.5
0.95	52.5

Plot your results on a graph

Can you describe the pattern in your results? *Open a spreadsheet program and follow the instructions in the panel below to draw a graph of pressure against volume.*

Does your pattern show a straight line?

Decide whether it would be possible to squash a gas down to no (or zero) volume. *Look at your graph*

Decide whether it would be possible for a gas to produce no (or zero) pressure.

Boyle found that you can get a better pattern if you plot the pressure against the **reciprocal** of the volume or 1/volume. *Follow the instructions in the panel to plot a new graph of pressure against 1/volume.*

What is the shape of this new graph?

How is pressure related to the volume - is it related **directly** or **inversely**?

Finding a constant

Boyle did some more calculations - he multiplied the pressure values by the volume for each of his results. *Follow the instructions in the final panel to get the spreadsheet to do this for you.*

What do you notice about your figures for the pressure multiplied by the volume?

Reference

With thanks to Linda Webb, Homerton College, Cambridge.

	A	B	C	D	E	F	G	H	I	J
1	Pressure	Volume	1/Volume	PxV						
2	x100,000N/m2	cm3								
3	3.13	16.0	0.063	50						
4	2.97	17.0	0.059	50						
5	2.78									
6	2.56									
7	2.42									
8	2.21									
9	1.97									
10	1.83									
11	1.60									
12	1.40									
13	1.20									
14	0.95									
15										
16										
17										
18										
19										

How to draw a graph with your results
1. Copy rows 1 and 2 as shown here.
2. Enter your results in columns A and B
3. Highlight all your results, e.g. from cell A3 to B14.
4. Get the program to draw a scattergraph.

How to get a better graph pattern
1. Move to cell C3 and enter the formula =1/B3
This works out 1 divided by the volume.
2. Copy cell C3. Paste it into cells C4 to C14
3. Highlight all your results in column A. E.g. from cell A3 to A14.
4. Hold down the CTRL or similar key.
5. Highlight all your calculations in column C. E.g. from cell C3 to C14.
6. Get the program to draw a scattergraph.

Extra
How to multiply the pressure by volume
1. Move to cell D3 and enter the formula =A3*B3
2. Copy cell D3. Paste it into cells D4 to D14

IT tools

Section

2

What this is about

Using a spreadsheet to 'model' an energy saving situation.

In winter, our homes lose heat energy...

- Heat is lost through the walls and floors through conduction.
- Heat is lost through the windows by conduction and radiation.
- Heat is lost through the roof by convection and conduction.

This diagram shows the percentage of home heating lost by these routes.

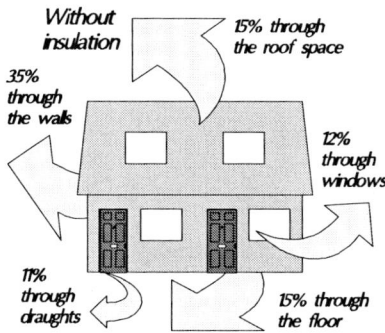

This diagram shows the percentage we can reduce these losses to by using various kinds of insulation.

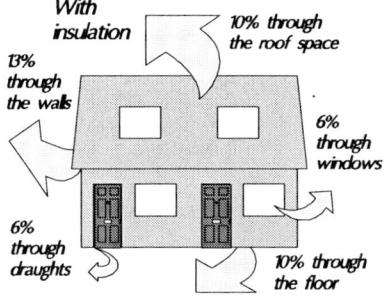

Choosing insulation

Open your spreadsheet with a new sheet. Use the diagrams to help you complete it.

1. Copy the headings in row 1.
2. Record the five routes of **heat loss** in column A.
3. Record the ways to **insulate** in column B, matching them up with column A.
   Carpet/underlay
   Cavity wall fill   Draught strip
   Double glazing   Loft insulation

Working out what heat loss costs

4. Record the Heat loss **without** insulation.
5. Record the Heat loss **with** insulation.
6. Our home energy bill - for a medium sized house was £400 last year. Record your own energy bill.
7. Calculate the percentage **Heat saving**.
9. Calculate the **Money saved each year**.

Questions

Which part of an insulated house loses the most energy?

Which method of insulation might save you the most money?

Extra

How much would you save in a year, if you insulated your home using all of the methods here? *Work out the sum of column F to find out.*

Which methods of insulation are the most important? *Draw a pie chart or bar chart to find out.*

Reference

Source of data: The Sciences for GCSE Activity sheets.

	A	B	C	D	E	F
1	**Heat loss**	**Insulation**	**Heat loss without %**	**Heat loss with %**	**Heat saving**	**Money saved each year**
2	Through floor	Carpet/underlay	15%	10%	5%	£20
3	Through walls					
4	Through window glass					
5	Through draughts					
6	Through roof space					
7		**Energy bill**	**£400**			
8						
9		*How to set up your spreadsheet*				
10		*1. Copy the headings in row 1.*				
11		*2. Enter five routes of heat loss into column A.*				
12		*3. Enter five ways to prevent heat loss in column B.*				
13		*4. Use the picture to add the Heat loss without insulation figures to column C.*				
14		*5. Use the picture to add the Heat loss with insulation figures to column D.*				
15		*6. Find out your heating bill for the year. Enter this in cell C7*				
16		*7. Move to cell E2. Type in the formula =C2-D2.*				
17		*8. Copy cell E2. Paste it into cells E3 to E6.*				
18		*9. Move to cell F2. Type in the formula =E2*C7.*				
19		*!0. Copy cell F2. Paste it into cells F3 to F6.*				
20						

IT tools

Section

2

Energy: home insulation II

What this is about

Using a spreadsheet to 'model' an energy saving situation.

We can reduce the amount of heat our home loses using insulation.

There are different kinds of insulation for different parts of the house.

- Cavity wall insulation fills the gap within the outside walls of some houses.
- Carpet and underlay insulates the floor.
- Foam draught strips fill gaps in windows and door frames.
- Loft insulation felt lies as a blanket in the roof space.
- Double glazing reduces heat loss through a single pane of glass.

It will cost you money to install any of these insulation methods. How much it will cost depends on how big your home is. Would insulation pay for itself by saving energy? By using a spreadsheet, you can find some answers.

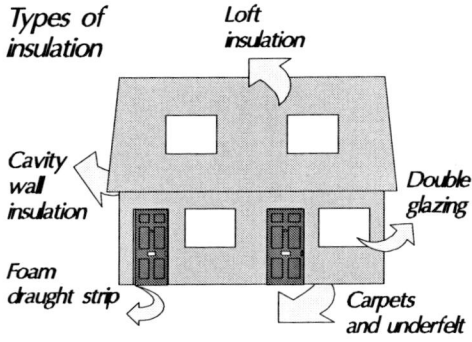

Types of insulation
Loft insulation
Cavity wall insulation
Double glazing
Foam draught strip
Carpets and underfelt

Information to collect

1. Try to find out your heating bill for the year. If you cannot, you can use our figure of £400 for a medium sized house.

2. Measure and work out the following sizes for where you live (or use our typical figures):

Part of house	Area in m or m2	Typical figures
Floor area		40
Outside wall area		200
Window area		10
Door and window frame	m	30 m
Ceiling area		40

Calculating the cost

Start your spreadsheet with the file **Keep warm** you have used in a previous exercise. The diagram below will show you how to do the following:

- Enter your data for the size of your home.
- Enter the data for the cost of the insulation per m.
- Work out the cost to install the insulation.
- Work out how long the insulation will take to pay for itself.

Questions

Which insulation is the most expensive to install?

Which insulation is the least expensive to install?

Which insulation might save you the most money?

Which insulation might save you the least money?

Jo said that 'Insulation is good for the environment but too expensive'. What do you think?

Extra

In a new block of flats, Helen chose the basement flat and Jo chose the top floor flat. Who made the best choice? **Enter your own flat sizes into your spreadsheet to find out.**

Reference

Source of data: The Sciences for GCSE Activity sheets.

IT tools

Section 2

	A	B	C	D	E	F	G	H	I	J
1	Heat loss	Insulation	Heat loss without %	Heat loss with %	Heat saving	Money saved each year	Areas of the house m^2	Cost of insulation per m or m^2	Cost to install	No of years to pay for itself
2	Through floor	Carpet/underlay	15%	10%	5%	£20	40	£17.00	£680	34
3	Through walls						200	£4.00	£800	9
4	Through window glass						10	£100.00	£1,000	42
5	Through draughts						30	£0.40	£12	1
6	Through roof space						40	£4.00	£160	2
7		Energy bill	£400							
8										
9		**How to set up your spreadsheet**								
10		1. Use the spreadsheet 'Keep warm'.								
11		2. Enter your sizes for parts of the house in column G.								
12		3. Copy the cost of the insulation in column H.								
13		4. Work out the cost to install the insulation as follows:								
		Move to cell I2. Type in the formula =G2*H2. Copy cell I2. Paste it into cells I3 to I6								
14		5. Work out how long the insulation would take to pay for itself as follows:								
15		In cell J2, type the formula =I2/F2. Copy cell J2. Paste it into cells J3 to J6.								
16										
17										

46

Gravity in the solar system

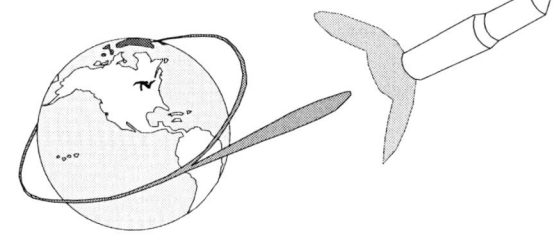

What this is about

An introduction to using a spreadsheet to do a calculation, sort a list and produce a graph. You may also use a word processor to write up the sheet and include the spreadsheet table and graph in it.

If you were to weigh yourself in space you would weigh nothing. There is little gravity pull on you and little weight to be measured.

If you could visit the moon or the planets it would be another story - all these places have a large gravity pull which gives you your weight.

Different places have different gravity pulls - just how different can be seen by working out your weight in each place.

What to do

Open a new spreadsheet and follow the instructions in the diagram below.

Sort your spreadsheet into order

Draw a graph with your results.

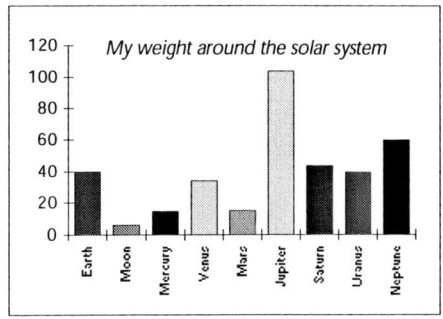

Questions

1. On which celestial body would you weigh the most?

2. On which celestial body would you weigh the least?

3. On which planet would you weigh the least?

4. If your weight doubled on earth, would you weigh twice as much on the moon? *Use your spreadsheet to find out.*

Extra

Include your spreadsheet table and graph with your answers. *Copy and paste your sheet and graph into a word processor if you have one.*

Note

*Celestial body: we can't use the word planet here because we have included the moon - which is not a planet.

	A	B	C	D	E	F	G	H	I	J	K	L
1	How much would you weigh on											
2	Body	Gravity	Your weight									
3	Earth	1	40									
4	Moon	0.16	6.4									
5	Mercury	0.37	14.8									
6	Venus	0.86	34.4									
7	Mars	0.38	15.2									
8	Jupiter	2.6	104									
9	Saturn	1.1	44									
10	Uranus	1	40									
11	Neptune	1.5	60									
12												
13	MAXIMUM		104									
14	MINIMUM		6.4									
15												

How to set up your spreadsheet
1. Copy rows 1 and 2 as shown here.
2. Copy columns A and B also as shown here.
3. Move to cell C3 and enter your weight.
4. Move to cell C4 and enter the formula =C3*B4
This works out your weight on the moon.
5. Copy cell C4. Paste it into cells C5 to C11
6. Move to cell C13 and enter the formula =MAX(C3:C11)
This finds your highest weight.
6. Move to cell C14 and enter the formula =MIN(C3:C11)
This finds your lowest weight.

How to sort your spreadsheet
1. Highlight cells A4 to C11.
2. Find the Sort command.
3. Choose column C as the sort 'key'.

How to draw a graph with your spreadsheet
1. Highlight cells A3 to A11
2. Hold down CTRL other special key.
3. Highlight cells C3 to C11
4. Get the program to draw a bar graph.

IT tools

Section

2

Force, mass, acceleration: cars

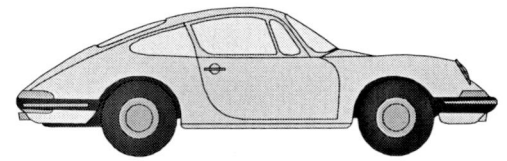

What this is about

Using a spreadsheet to handle data on cars. The program draws graphs and helps with calculations.

How would you choose a car? On price, quality, performance or running costs? To help you choose the car magazines publish data. We have put some data in a spreadsheet to help you decide.

Quick questions

Which car has the highest maximum speed?

Which car has the best acceleration?

Which car seems the least expensive to run?

Which car is the heaviest?

True or false?

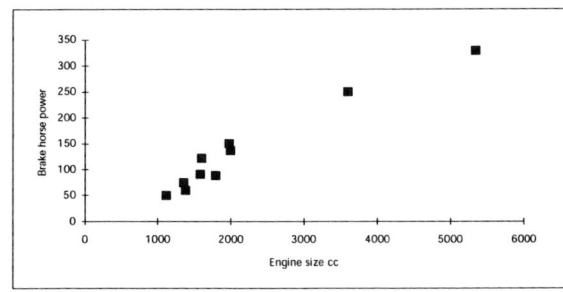

Draw scattergraphs, like this one, to see whether the following statements are true.

The car's maximum speed depends on how heavy it is.

The car's maximum speed depends on its 0-60 acceleration.

The car's maximum speed depends on its engine size

The car's acceleration affects its fuel consumption.

Convert the units to SI

Unit	Non-SI	SI
Mass	1 cwt	50.80 kg
Distance	1 mile	1609 m
Time	1 hour	3600 seconds
Acceleration	0-60 mph	m/s^2
Torque	1 ft.lb-wt	1.36 newton.m

The data you have is non-SI. Use this **Conversion table** to convert the units for Car mass into kg, for Maximum speed into m/s, and for Torque into Newton.metres. **Add three extra columns to your spreadsheet. Then use formulae, to convert the units as shown here. The first row has been done for you.**

Car mass	Maximum speed	Torque
kg	m/s	N.metres
=D4*50.8	=G4*1609/3	=O4*1.36

Choosing a car

Which car would you choose:
a)   For town driving
b)   For long distance driving

Serious maths

The relationship between force, mass and acceleration is given as F=ma. Does this relationship hold true for the cars in the table? (Torque is the maximum driving force of a car). How might friction affect your answer?

The quoted torque only applies at one particular engine speed (usually about 2000 rpm) so for a real acceleration in which the engine speed is changing, the available torque will also change making the average over the period of acceleration different from the quoted value.

Reference
With thanks to Linda Webb, Homerton College, Cambridge.

IT tools

Section **2**

	A	B	C	D	E	F	G	H	I	J	K	L	M	N	O
1							Data on Cars								
2	manufacturer	model	list price £	kerb weight	country of origin	cost per mile p	maximum speed	0-60 mph seconds	fuel consumption				engine size cc	Bhp	Torque
3				cwt			mph		urban	56mph	75mph	touring	cc		ftlb
4	Aston Martin	Virage	133574	37.8	GB	n/a	155	6.5	12.1	25.8	20.4	17.6	5340	330	350
5	Citroen	ZX1.4	9680	18.7	Spain	28	104	12.4	33.2	50.4	39.2	39.0	1360	75	82
6	Fiat	Uno 1.1	7095	15.2	Italy	25	93	16.0	36.7	58.9	43.5	44.0	1108	50	62
7	Ford	Mondeo 2.0	14095	24.6	Belgium	40	130	9.5	25.2	44.8	36.2	32.9	1989	136	133
8	Peugeot	405 1.6i	10170	20.9	GB	34	109	14.3	28.8	48.7	36.2	35.6	1580	90	98
9	Porsche	Carrera 2 Coupe	50450	26.6	Germany	139	159	5.1	16.5	36.2	29.1	24.6	3600	250	220
10	Rover	216 Cabriolet	15645	22.3	GB	46	114	9.7	28.2	44.3	33.8	33.6	1590	122	103
11	Vauxhall	Corsa 1.4i merit	7605	17.4	Europe	26	91	17.0	32.1	47.1	36.2	36.9	1389	60	76
12	Volkswagen	Golf GTi	15999	22.9	Germany	44	132	7.9	26.2	44.8	35.8	33.3	1984	150	133
13	Volvo	440Li	10995	19.9	Holland	37	109	12.0	26.2	50.4	39.2	39.2	1794	89	103
14	Kerb weight is car empty with full tank measured in cwt						Data: "What Car?" magazine. All the units of measurement are non SI								
15	The touring figure for fuel consumption is 50% urban +25% @ 56mph + 25% @ 75mph figures						Cost per mile is calculated over 3 years and 36000 miles								

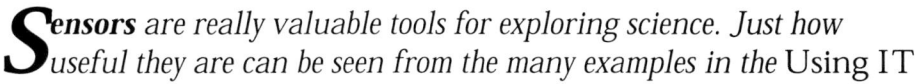

Sensors *are really valuable tools for exploring science. Just how useful they are can be seen from the many examples in the* Using IT *section of this book. Each idea can offer something towards giving pupils a better understanding of science. They measure fast changes and they measure with precision. They extend the range of things we can do from timing a falling mass to measuring our pulse. Most importantly, they provide pupils with a better feel for the changes they measure.*

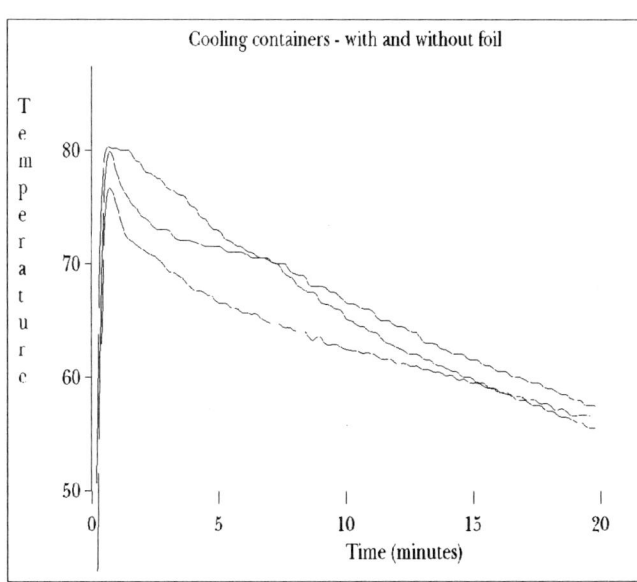

The scope for using sensors in science is so huge it merits a book* in itself. However, two interesting exercises using sensors follow. Both are technically easy but the 'science' is anything but superficial.

*Two companion and complementary books to this look exclusively at the uses of sensors and control technology in science: **The IT in Science book of Data logging and Control.** *ISBN 0 9520257 1 X and* **Data logging in Practice** *ISBN 0 9520257 4 4. The first title concerns the opportunities and scope,the second looks at ways to implement the technology in school.*

About the worksheets that follow

Insulation: wrapping food in foil - page 51

Pupils measure temperatures as beakers of hot water cool. The beakers can be wrapped both loosely and tightly in aluminium foil to see if this makes a difference. The activity, which will raise questions on methodology, is written as an investigation. The graph *(above right)* shows a possible result: there were three beakers, one wrapped tightly, one wrapped loosely and one not wrapped at all.

The pupils can go on to see if it matters whether the foil is shiny side out or in. They will need to improve their technique to perfection. (We were disappointed to find that the difference was negligible.). *IT Level: easy*

Insulation: keeping houses warm - page 52

Pupils set up a series of model houses *(from ASE)* and use temperature sensors to see how well they retain heat. They can heat the houses by using a microscope lamp or by using a calorimeter of hot water. The pupils will need to measure for some minutes before the houses reach a steady temperature. If they use the microscope lamp they will need to switch it off at this point. You should discuss how representative the findings are.

They can go on to compare a detached house, semi-detached house, and a block of flats. They might do this by measuring the heat loss from houses which are alone, side-by-side or stacked on top of each other. IT *level: easy*

IT tools

Section

2

Data logging and control glossary

Analogue Port - a socket on a computer which you can connect analogue sensors to. Some computers have an analogue port build in to them, others have this as an option. On modern sensor systems you connect the sensors to an interface which in turn plugs into the serial port.

Analogue sensor - a sensor which has many states and can provide readings over a wide range of change.

Analogue to digital converter - part of a computer interface which converts an analogue reading from a sensor into a digital reading which the computer can interpret.

Control box - an interface which allows you to switch and power lights, motors and buzzers. The box will have inputs for digital sensors - and may take analogue sensors.

Control language - you use a control language to write programs for control systems.

Control software - the programs used to read information from sensors and switch devices on and off.

Data logging - collecting data from sensors. To do this away from the computer you need a 'data logger'.

Data logger - a self-contained device to collect readings from sensors away from the computer. When all the readings have been taken you connect the data logger to the computer to transfer the readings.

Data logging software - software which is designed to record and display the readings from sensors. Usually supplied specifically for your data logging kit.

Digital sensor - a sensor or switch which has two states, on or off.

Interface - a device you use to connect the sensors to the computer.

Light switch - a digital light sensor which responds to something covering it. You can use one to sense when dusk occurs.

Position sensor - measures the angle of movement.

Pressure mat - a switch that responds to momentary pressure. Put under a mat as part of a burglar alarm system.

Proximity switch - a switch that responds when close to another object. One type of proximity switch is a reed switch which is triggered when brought close to a magnet.

Push Switch - a switch that responds to momentary pressure. Use as a bell push or to control a pelican crossing.

Rotation sensor - measures the speed of rotation. Use for monitoring the wind speed, the speed of a motor or gears.

Serial Port - a computer socket where you may connect an interface. Sensors then connect to the interface.

Sensors - devices which respond to a change in the environment. There are as many as thirty different sensors available to schools.

Sound sensor - measures the sound level. Use to study sound travel and sound proofing. Sound is measured in decibels.

Sound switch - a digital sensor which responds to sound. Can be used to measure the speed of sound.

Temperature sensor - measures how hot something is. Use to study cooling, heating, insulation and the weather.

Time graph - a way of showing how sensor readings change over a period of time.

Timing light gates - a digital sensor which responds rapidly to changes in light level. Used for timing events with great accuracy.

Toggle switch - a type of digital sensor. It is a two-position switch similar to the switch used to turn a television on and off.

User Port - a socket on some computers and interfaces where you can connect control boxes.

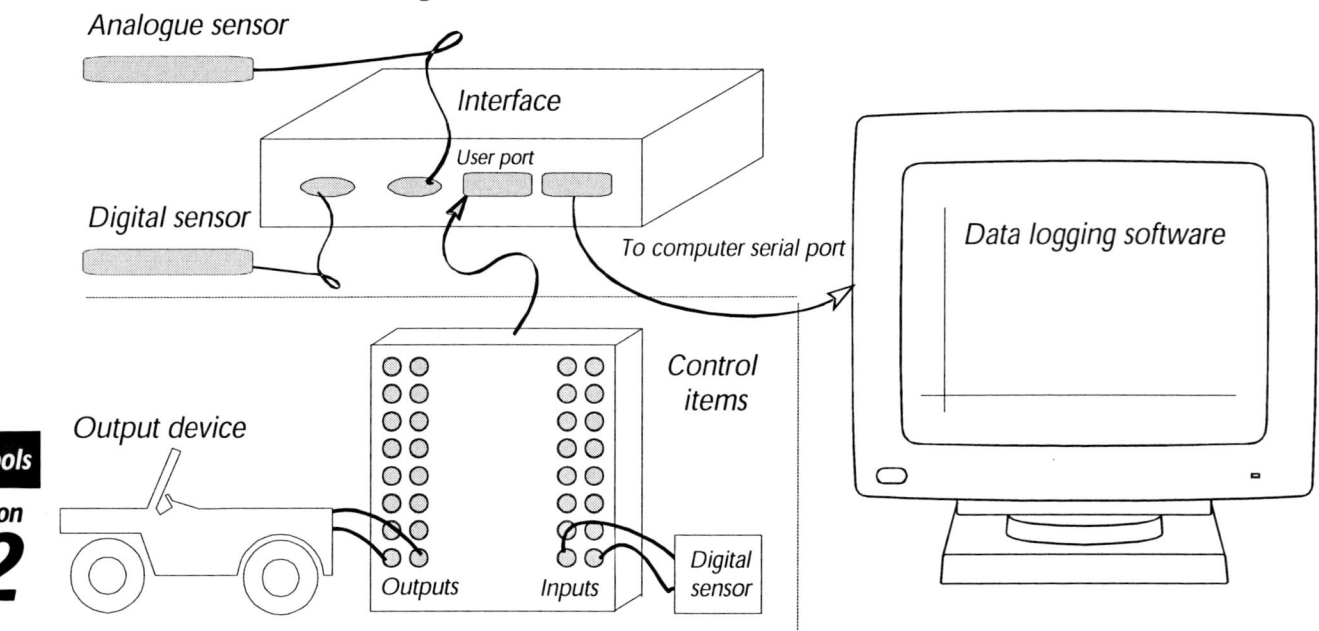

Analogue sensor

Interface

User port

Digital sensor

To computer serial port

Data logging software

Output device

Control items

Outputs

Inputs

Digital sensor

Insulation: wrapping food in foil

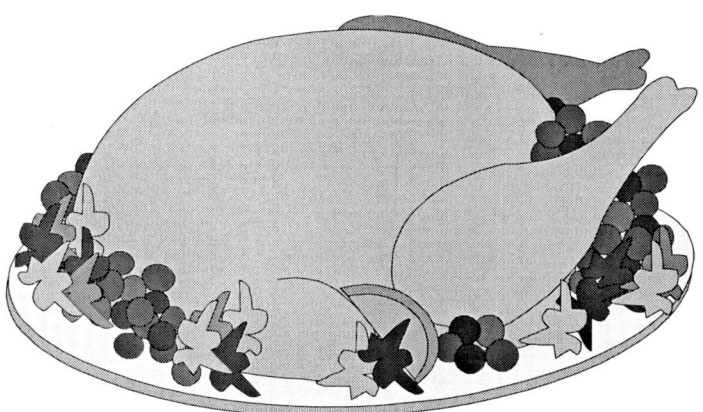

What this is about:

Using temperature sensors to measure changes in temperature over time.

In many recipes we wrap our food in aluminium foil. People say they do this 'to keep the flavour in' or 'to stop it drying out' or 'to make it cook more evenly' or even 'to keep the heat in'.
As a scientist, you could test any of these ideas - but in this investigation you will test the idea that foil keeps the heat in.

You will need

Beakers, foil, hot water, temperature sensors and computer system.

What to do

You are going to test the hypothesis, 'Foil helps to keep the heat in'

You will need two beakers - one will be wrapped loosely with foil and one will be unwrapped.

You can use temperature sensors to take two sets of readings at the same time. Briefly describe how you will do your experiment. Remember to say, when you will need to start measuring and when you will need to stop measuring.

What things will you try to keep the same to make your experiment a fair test?

You will obtain a graph with two temperature-time lines on it. Sketch how you think these lines will look.

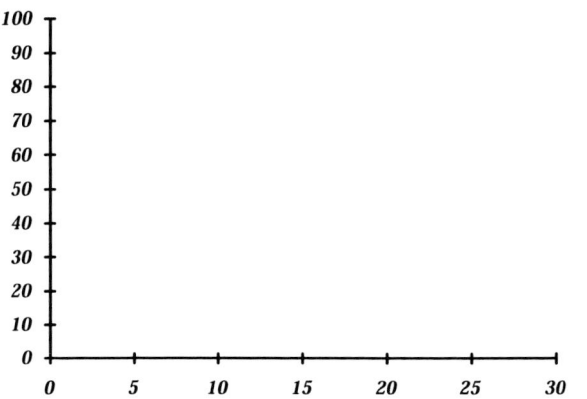

Using the computer

Connect your sensors to the interface.

Connect the interface to the computer.

Open your sensing software.

Set up the software to measure temperature.

Check that the software will record for long enough.

Remember to start the computer recording.

Watch the graph during the experiment to make sure things are going according to plan.

Questions

How does the graph show you that using foil makes a difference? *Zoom or change the axes on the graph if this makes it clearer.*

Does foil keep the heat in?

More to explore

- People also say that to keep the heat in you must wrap the food loosely with foil. They say that if you wrap it tightly it will not work. Is this so?

- Some people cook their food in foil with the shiny side-in. Is this a good idea?

- Astronauts wear shiny suits. Would an astronaut get 'cooked' in the heat of the sun? If you had to advise the designer of their space suits, what would you say?

IT tools

Section

2

Insulation: keeping the house warm

What this is about:

Using temperature sensors to measure changes in temperature over time.

Insulating a house prevents heat loss. In this activity you will investigate different ways of doing this.

You will need

Model houses, insulating material, polystyrene, acetate film, masking tape, temperature sensors and computer system. Lamp and power supply, beakers, calorimeters and hot water.

What to do

Decide how you will heat your house. You can use either: a lamp or a container of hot water

Then set up the four houses shown.

Starting your experiment

Connect your sensors to the interface and the interface to the computer.

Set up your sensing software to measure temperature.

Get the software recording for the next 30 minutes.

Heat the houses until the temperature is steady. Then allow them to cool.

Single glazed Loft uninsulated

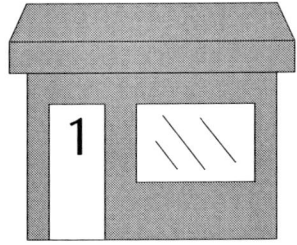

Double glazed Loft uninsulated

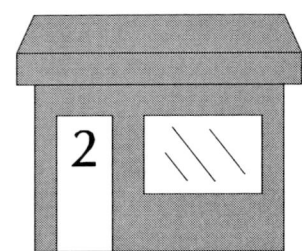

Single glazed Loft insulated

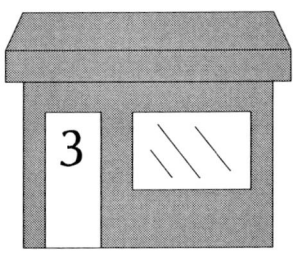

Single glazed Wall insulated

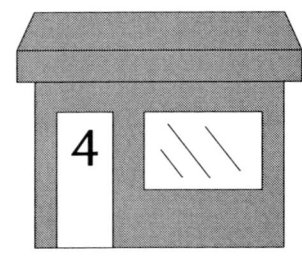

Questions

To record what you did, copy and fill in the table below.

Include the graphs in your report.

- How does the graph show you that double glazing makes a difference?
- How does the graph show you that loft insulation makes a difference?
- How does the graph show you that wall insulation makes a difference?

Extra

Investigate which of the following would keep you warmest:

- A detached or a semi-detached house.
- A top flat or a basement flat.

IT tools

Section 2

			Loft insulation	Wall insulation	Window insulation
1	House		None		
2	House				
3	House				
4	House		None	Wall	

W**ord processing* is more than just typing - it is a way of improving the quality of written work. It allows pupils to jot ideas on-screen and develop them. It allows them to re-think finished work and even then refine it. ***Because *word processing allows them to improve their work, they do. For as long as writing is part of learning in science, the need for pupils to have access to this powerful technology will remain.*

There is a hidden bonus too. With a computer screen as their focus, pupils can work together and discuss their task more freely than they could before. Pen and paper certainly have their role, but as the medium for collaborative work the word processor excels.

Pupils have a lot to document in science. They would be unusually fortunate to have sufficient access to computers to do this on a computer. The ideas in this book take this into account. They suggest a 'quality not quantity' approach - making the occasional use of the word processor into a task which has a real purpose, is reflective or is more alive.

For example, the ideas suggest pupils write leaflets, advertisements, newsletters and stories. They can make a poster to stop the waste of energy or write a Which *report to compare different metals. They might 'just' plan an experiment or 'just' write a 'lab report'. But, even here the advice is to prime the computer with key questions: What were you trying to find out? Which results are useful? Why do you think that happened? What have you leant from this?'*

In the Using IT section you will find many more such examples. In the rest of this section you will find a number of interesting ways of using a word processor.

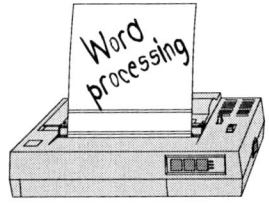

IT tools

Section

2

Word processing glossary

Box - a rather ugly way of emphasizing headings.

Block - a section of writing you have selected to format or move elsewhere.

Centred - where you put the writing exactly in the middle of a line. Useful for headings and sub headings.

Change (Replace) - a feature which allows you to change any word or phrase to another. For example you can use it to change the name of a person called Fred to Frederick, all the way through your work. A useful feature but only in a longer piece of writing.

Copy - where you can copy a section of writing in the work to save you retyping it.

Cut - where you can remove a section of writing from the work. You may paste it back elsewhere in the work or even paste it into another program.

Find (Search) - to find a word or phrase. Useful for finding your place in a longer piece of writing.

Font - a feature which allows you to change the style of the letters from say, plain to decorative.

Format - where you can change the letter size, style or font.

Object - a strange but useful feature where you can place a photo, picture or graph on your typed page.

Paste - where you can replace a previously cut or copied section of writing back into the work.

Select - where you can choose a section of writing. For example, you might select a sentence to put it in a bolder type.

Size - where you can change the size of the lettering from small to large - for example, in headlines.

Style - where you can change the lettering style to bold, italic or bold and italic.

Tab - a special key on the keyboard which inserts a long space. It allows you to line up columns of words or numbers although the 'Table' feature does this better.

Underline - a rather dated way of emphasising headings and side headings.

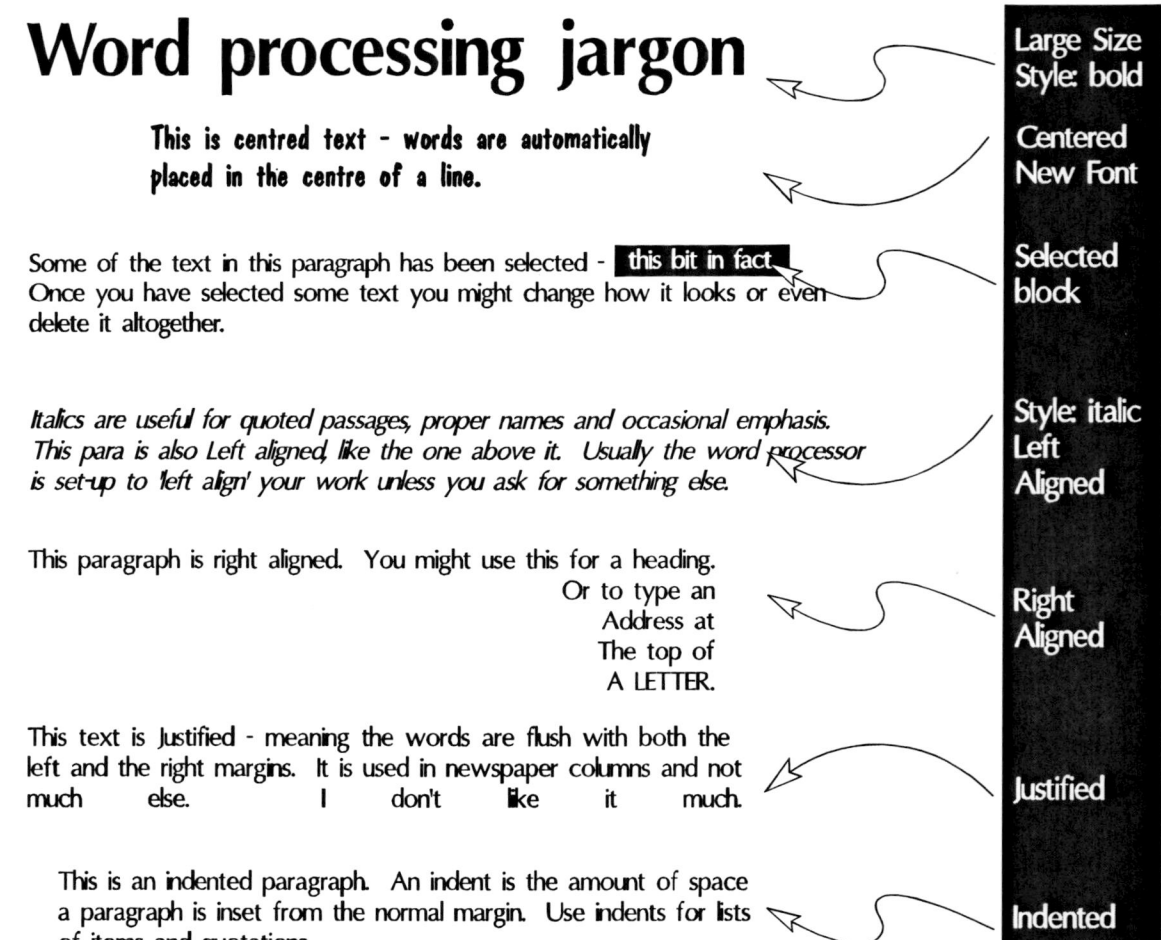

IT tools

Section

2

Word processing ideas

An experiment planner - page 57

It is quite daunting to have a blank sheet of paper and be asked to plan an experiment. This situation clearly needs a stronger focus and it needs breaking down too. In the exercise on this sheet the pupils use the word processor to help them plan a science investigation.

You need to provide the pupils with a copy of the page as a word processor file. Type the page into the word processor - edit the questions if you wish, then save the 'worksheet' on the disk.

Aim to get pupils drafting and revising their plan rather than just typing it nicely. They tend to spend more time preening it than thinking about science. For example, discourage the pupils from formatting their work until they have finished it.

You may be able to set the file to 'read only' so that your original copy of the file remains intact. In time you will want to amend this planning sheet, perhaps to direct it to a particular group of pupils. Even so, this worksheet could be the most flexible and useful worksheet you ever type into a computer!

Separating salt from rock salt - page 58

This is the word processor equal to a paper cut-out activity but this one requires very little time. The pupils are given a set of steps in the wrong order and they have to sort them out.

First check that your word processor will allow you to move phrases from one place to another. Next, open a new file and type in the phrases on the worksheet provided here. Save the file and distribute it round the class. Finally, get the pupils to re-order the sentences. The example here concerns the procedure for separating salt from rock salt.

The work is extended by a second set of phrases which explain the reasons for each step of the procedure. The pupils have to move these to the right places too.

Other suitable contexts include the steps of a 'testing leaves for starch' experiment, the water cycle, the life stages of the butterfly or how to identify an unknown chemical.

Making posters and advertisements

Pupils can use a word processor to prepare a poster on the moon, space travel, safety at home, healthy eating, keeping fit and pollution.

Your holiday in the stars!

They might for example, do a data search on aluminium metal and then prepare a poster advertising its features.

You will need a word processor that lets you include pictures and gives large printed type. Graphics programs may also be used here.

Writing reports

Pupils can write up their experimental work with a word processor. Most programs will allow them to add graphs, and pictures to it. Many will even allow them to illustrate their work with clips of video!

Experiment Report

Writing stories

As a follow-up to a block of work, the pupils can use the word processor to develop a story. For example, they might summarise some work on plants and photosynthesis by completing a story which begins, "One day Jan woke to find the sky was dark..."

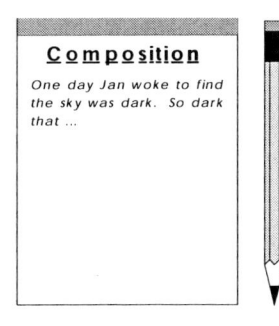

Composition

One day Jan woke to find the sky was dark. So dark that ...

IT tools

Section

2

Word processing ideas

Missing words exercise

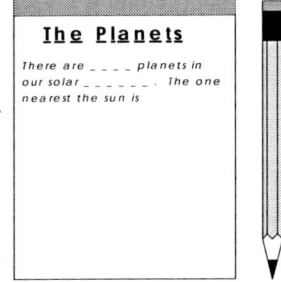

You can use the word processor to provide the computer equivalent of a 'directed text reading activity'. The most common example of this is the missing words exercise. The activity is a useful means of focusing attention on a passage of writing.

Using your word processor you type in the text, remove words and leave spaces for them. As this is meant to be an exercise, rather than a test, avoid removing too many words at the start of the text. This way it retains sufficient clues to allow the pupils to get a feel for the context.

You can use programs such as **Developing Tray** - which will remove letters, at random, from a piece of text. The program is worth exploring as it provides more hints and help than your word processor can. The idea can be very educational but how well it works depends on your choice of text (keep it rich and short) and how well pupils work together.

Overlay keyboards

An overlay (or *Concept*) keyboard is a flat A3 or A4 tablet which plugs into the computer. It is an alternative to the QWERTY keyboard. Instead of letters on key tops you have symbols, pictures or even words. To use it you place a sheet of paper on the tablet surface and mark areas with symbols, pictures and words. You then use a special program to attach words to the areas on the keyboard. When the pupil presses each area on the tablet, words are typed into the computer.

The keyboard overlay can be a memory jogger and a great help with spelling. In this way the keyboard makes word processing more accessible to younger as well as special needs children. The keyboard has many uses across the school and several finished examples are given in *Supporting Science* - an IT resource with a special needs focus from the *NCET*.

More recently 'software' keyboards have appeared - which save you the difficulty of connecting things up by putting a 'keyboard' grid on the screen. So when you click on word or pictures in the screen grid, words appear in your word processor. An excellent example of this is Crick Computing's **Clicker** program (PC / Arc - SEMERC).

Making a newspaper

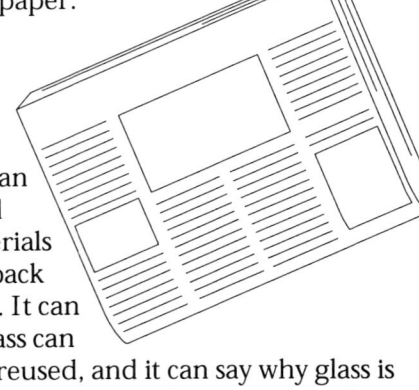

A newspaper project can provide an excellent focus for science work. For example, the pupils might prepare a science magazine or a recycling campaign newspaper.

The recycling newspaper can talk about the kinds of materials that can be recycled and how these materials find their way back into the system. It can explain how glass can be melted and reused, and it can say why glass is sorted into different colours. It can explain the problem caused when bottle caps get thrown into the bottle bank. It might talk about the quality of recycled paper and the uses of recycled paper. It can be illustrated with graphs showing the growth of recycling and pictures showing how much waste one household produces in one year.

The class science magazine can feature the discoveries and writings of members of the class. They can add pictures of themselves and talk about the investigations they did. The pupils can tell the rest of the school about their noise survey. They can tell others about their efforts, scientific that is, to find the best brand of trainers.

There is plenty of scope here for a whole class project. If it seems appropriate, you can have teams of picture editors, reporters, sub-editors, printers and so on, all working to an agreed production schedule.

You need a program that lets you use pictures and gives a decent print-out. You may also need more time than you can afford, but the evidence to date seems to indicate that children benefit from such activity enormously.

Connect up to the world

And why not consider the possibilities for sharing all this work with others, even on the other side of the world? By using a modem and a phone line and subscribing to an **Internet on-line service**, pupils can share their work or start up a dialogue with others. They might post a question to a scientist or post their work for the world to see. Don't underestimate the potential of this for injecting a 'real-life' element.

IT tools

Section

2

Planning an experiment

What this is about

This is about using a word processor to plan an experiment or investigation. The text below should have been typed into a word processor for you.

Being able to plan an experiment is an important skill for a scientist. To make the job easier we have given you a set of instructions.

This word processor file will allow you to plan your experiment. Complete the sentences below, in any order. Under each are some typical answers.

Do not 'format' your work until you have finishing all your writing.

Topic

Type in the name of the topic as your heading.

Hypothesis

We are trying to find out if ...

When you complete this you are making a hypothesis. For example, your hypothesis might be that ...... *the resistance of a wire changes with its length.*

Prediction

What I think will happen is ...

When you complete this you are making a prediction. For example, you might predict that ...*the resistance of a wire increases with its length*

I think this will happen because ...

You need to explain yourself. Scientists always explain why, even when they make guesses. For example,
... *the longer the wire is, the harder it is for electricity to get through it.*

Apparatus

I will need the following equipment ...

You need to list the apparatus you will need. You might fill this in after you have done the next couple of questions.
... *Constantin wire, ruler, power supply, voltmeter, ammeter, connecting wires.*

Proposed method

What I plan to do is ...

You need to describe what you will do in your experiment. This is your experiment method.
...

What I will measure or look for is ...

You need to measure the variable you think will change. This, by the way, is the dependant variable. The clue to what to measure may be in your prediction or hypothesis. For example, you might say,
... *the length of the wire, the voltage and the current. I will have to work out the resistance using Ohm's law.*

I will take readings ... using ...

You need to work out how many readings you need to take and how frequently you will take them. For example, you might take a reading,
... *for every 10 cm of wire ... using meters.*

In my experiment, I will change...

What you change in your experiment is called the independent variable. For example, you might change
... *the length of the wire.*

In my experiment, I will keep the following things the same ...

To make your experiment a fair test you will need to keep control of things. For example,
... *I will keep the rest of the circuit exactly the same.*

To be safe I will make sure that ...

You need to make sure no one gets hurt and nothing is damaged. Say what precautions you will take.

Now read through your work. Format the text and delete the help sections.

Note to teacher

Provide the pupils with this 'computer' worksheet - which you will need to type into the word processor in advance. Save the 'worksheet' on the disk and set the file to 'read only'.

IT tools

Section

2

Separating salt from rock salt

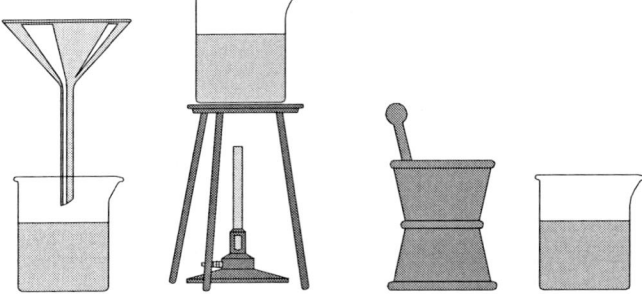

What this is about

This is about using a word processor to put sentences in the correct order. The text following this paragraph will need to be typed into a word processor for the student.

You will have done an experiment where you tried to obtain pure salt crystals from crushed rock salt. The experiment works because salt dissolves in water and the sand in the rock is insoluble. Here is how you might have done the experiment, but the steps are in the wrong order. Can you sort them into a correct and useful order?

What to do

Use this word processor program to arrange the steps below in the correct order. Keep each step on its own line. You can use the copy and paste commands to do this.

Step
Add some hot tap water to the solid...
Discard the residue...
Filter off the salt crystals...
Filter the mixture...
Grind the mixture into a powder...
Heat the filtrate to evaporate some water
Leave the solution for a while...
Stir the mixture and water...
Tip the powder into a beaker...
Wash the residue with a little water...

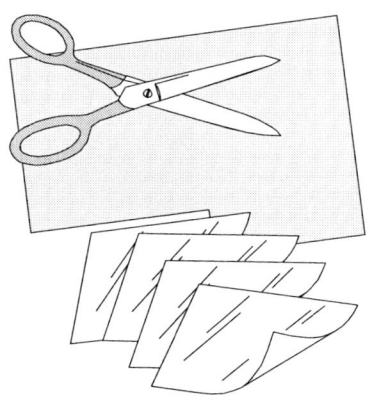

Extra

Use the copy and paste commands again. This time you must add reasons to each of the sentences you have just rearranged. Just add the 'reasons' below to the end of each step. For example,

*Add some hot tap water to the solid...**so that the salt dissolves quickly**.*

Reason
...because it is just sand.
...because you will need to keep it in something suitable.
...so that the salt dissolves quickly.
...so that the salt dissolves quickly.
...so that the salt dissolves quickly.
...so that the solution becomes saturated.
...to let it cool and form crystals.
...to separate the crystals from the solution.
...to separate the sand from the salt solution.
...to wash through the last trace of salt.

Answers

Grind the mixture into a powder...so that the salt dissolves quickly.

Tip the powder into a beaker...because you will need to keep it in something suitable.

Add some hot tap water to the solid...so that the salt dissolves quickly.

Stir the mixture and water...so that the salt dissolves quickly.

Filter the mixture...to separate the sand from the salt solution.

Wash the residue with a little water...to wash through the last trace of salt.

Discard the residue...because it is now just sand.

Heat the filtrate to evaporate some water...so that the solution becomes saturated.

Leave the solution for a while...to let it cool and form crystals.

Filter off the salt crystals...to separate the crystals from the solution.

IT tools

Section

2

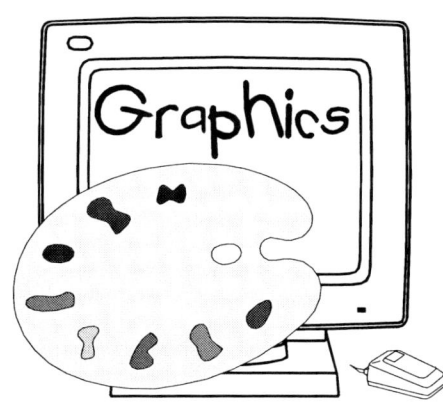

*G*raphics *is a pupil's first language. Tools which allow pupils to illustrate their investigations and observations can only help them to communicate better. It's hard to imagine a useful piece of scientific communication without graphics.*

Drawing programs *allow pupils to draw perfect lines, rectangles and circles. That alone makes them valuable for drawing diagrams. They also allow pupils to easily correct errors, make things bigger or smaller or copy a picture from one place to another. They can even build up a library of pictures and constantly recycle them. Such programs can save much time and effort.*

For some types of illustration, photographs are essential. There are various devices to get photographs into the computer. One is the **scanner** *- an affordable accessory which teachers and pupils will find many uses for. This allows you to use pictures in your work with ease. Another device is the* **digital camera** *which, as it uses magnetic disc rather than film, gives you instant photographs. This makes so many projects realistic it ought to be a resource available to the whole school. Pupils might use one to record their experiments or make a pictorial database of creatures they found on a field trip. You might use one to do time-lapse photography of a growing plant.*

Science is rich in processes that require moving images. Pupils can use a **video camera** *to record animals in action or to bring back evidence from field trips. You might even record pupils doing experiments, and ask them to comment. There is also* **multimedia** *authoring. With this pupils can use the computer to capture pictures, sounds and video. They can then assemble these to produce a very effective report of an event. Multimedia is finding many uses in industry and schools are finding that pupils are highly motivated to produce their offerings.*

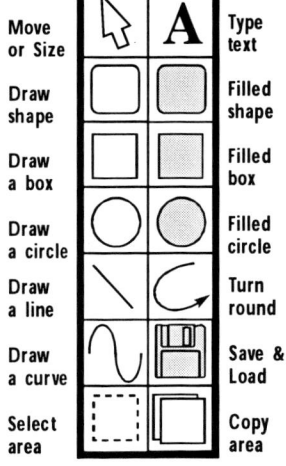

Move or Size	↖ / A	Type text
Draw shape		Filled shape
Draw a box		Filled box
Draw a circle		Filled circle
Draw a line		Turn round
Draw a curve		Save & Load
Select area		Copy area

About the body cut-out sheet

The body - page 60

Drawing programs can be used for picture cut-out activities. The potential for learning is not only as good as it ever was, but, from the pupils point of view, there can be a considerable time saving. From the teacher's point of view, the activity may take minutes or hours to set up - it just depends on the software.
IT Level: easy/medium

More cut-out ideas

· Arrange the phases of the moon into order.
· Arrange the steps of the water cycle.
· Arrange weather symbols on a map.
· Assemble the parts of a flower.
· Match pictures of animals to their names.
· Match pictures of materials to their uses.
· Place teeth in their places in the mouth.
· Sort materials into metal and non-metals.

IT tools

Section

2

What this is about

Learn about the sizes of parts of the body and how they fit together. You use a graphics program as if you were doing a cut-out exercise. The graphic below should have been prepared by the teacher and saved on disc for you.

Or get the MS Word file from www.rogerfrost.com

Can you fit each part of the body into its place?

What to do:

1. The organs are in the wrong boxes: put the pictures of the organs in their correct boxes.

2. The organs are the wrong sizes: scale each organ to fit its box

3. The organs are in the wrong places: put each body part into its proper place.

4. Label your diagram.

5. Delete the boxes

6. Print your body.

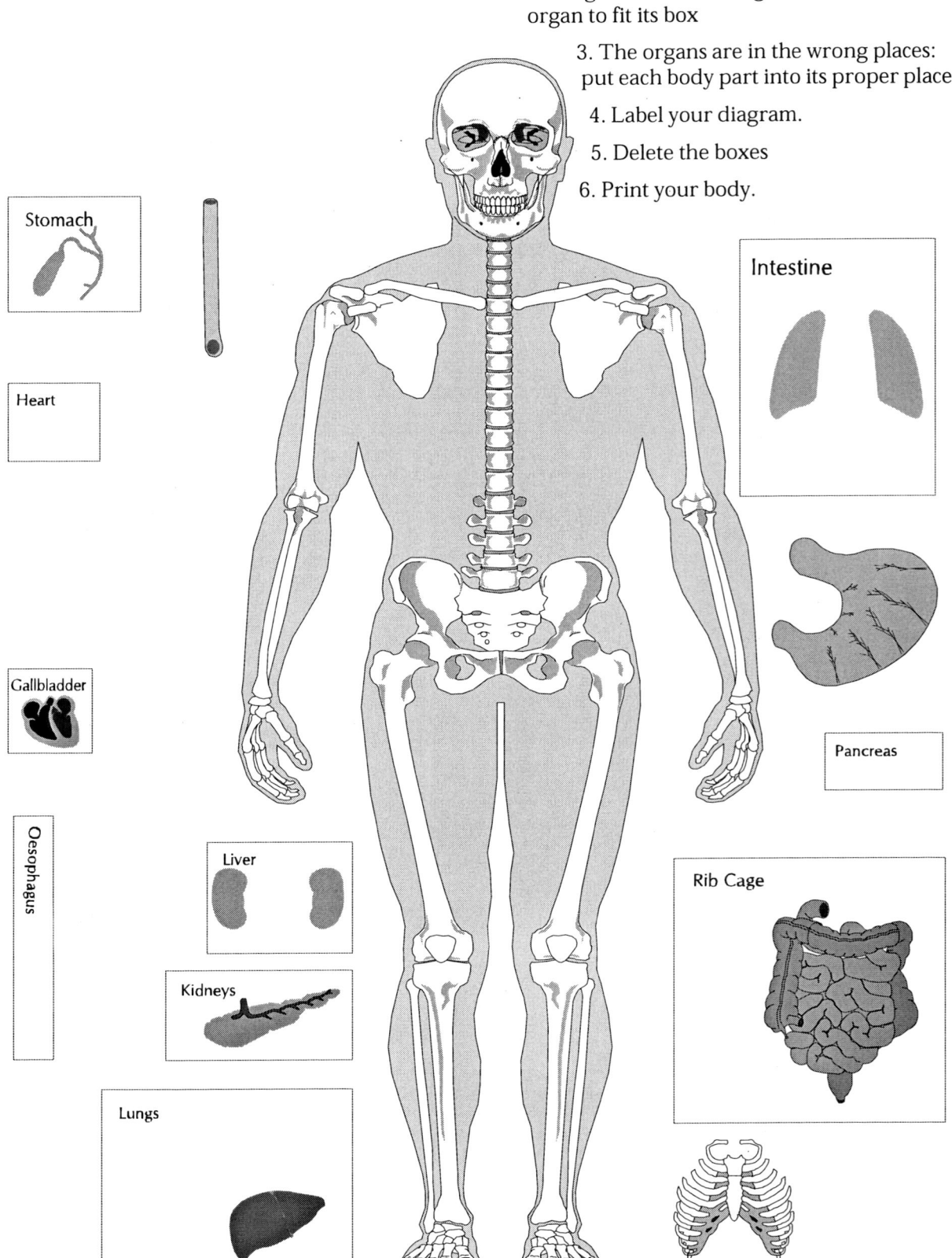

Stomach

Heart

Intestine

Gallbladder

Pancreas

Oesophagus

Liver

Rib Cage

Kidneys

Lungs

IT tools

Section

2

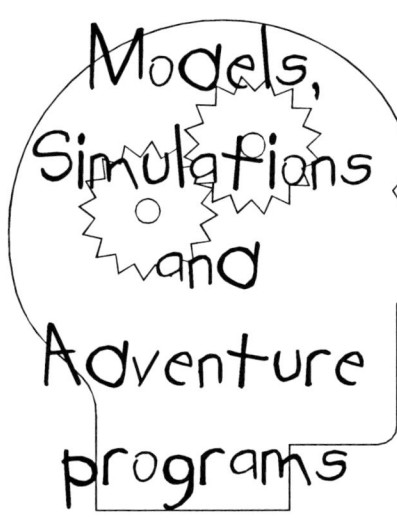

Software models

In science we try to understand the natural and physical world, by breaking it down into manageable parts. We build **models** to represent it.

A 'part' of the world might be the home, a forest, a seashore, or even some science idea like kinetic theory. And computer programs can help model these. They give us a chance to explore, to play with variables, to test ideas and gain an insight into how things tick.

Model builders, simulation programs and adventure games are the sorts of computer programs that allow us to do this. Each is a distinct type of program although that distinction is frequently blurred in practice.

Model builders and spreadsheet programs allow you to build, explore and to change models. Simulation and adventure programs merely allow you to explore.

Modelling is fascinating and thought-provoking. It can tax the brain heavily and may miss some pupils completely. But now and then you will find a piece of software which makes a difficult idea much more accessible.

What follows is an overview of the kinds of activities broadly termed using models. More can be found in the listings of the Using IT section of this book.

Modelling worksheets

IT tools

Section

2

Examples of models

An overlay keyboard together with a program such as *Touch Explorer*, is a very flexible tool to create modelling and mapping activities. You can place a map of the school environment on the keyboard. The map can show trees, green areas, and a pond. The computer program can then show messages on the screen when parts of the map are touched. The program lets you set this up with moderate ease.

Spreadsheets allow you to model using numbers. You can model your use of electricity at home, your use of water and energy too. For ready-made examples, see the *Essex Spreadsheets* package (Essex).

Pupils can collect and enter their heights, weights and other details to make a computer database - a model of the class. Using their data they can produce graphs to explore patterns and relationships in the data.

Fitting a power function to a graph of a cooling beaker of water is modelling. The human body cut-out on the previous page is a model. Building a robot buggy and getting the computer to control it is also modelling.

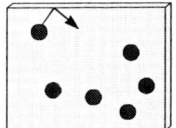

Moving Molecules was a classic illustration of kinetic theory in software

There are modelling programs on many aspects of science. Two examples included *Science Explorer* (Granada Learning) which lets you grow tomatoes under different conditions while *At Home in Wattville* lets you move around the house switching things on while you watch your electricity bill grow.

States of Matter (New Media) is a kinetic theory model - it shows us how temperature affects particles. *Power Package* (UE) took a look at electricity supply during periods of high and low demand. *Red Shift* (Maris) lets you explore a model of the solar system.

There are titles which simulate taking the human body apart, walking in space, building electrical circuits, or improving the environment. Really good examples are scarce - *Creatures* (Future Skill), *Crocodile Clips* (firm of same name) and *Edison* (Quickroute) are on-task. New Media have a wide range of neat simulations you could use too.

There are simulation programs to model photosynthesis, the control of blood sugar, chemical equilibria and the cardiovascular system. You can model waves and see how they interact. In fact, there are some very polished examples that are well worth seeing - in particular, the *Explorer Gateway* series (TAG) and *Simnerv* (see under Nerves) lets you stimulate a frog sciatic nerve.

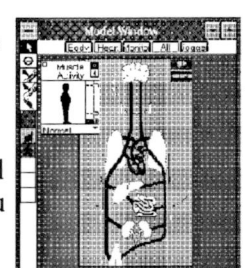

Model builders are generic programs which allow you to build your own models. You will find titles such as *Crocodile Physics* (PC / Mac - RM) - which lets you experiment with motion, lenses and gravity. And then there's *Interactive Physics* (PC / Mac - TAG) which lets you to model heat loss from a house, or the populations in an ecosystem.

	A	B	C	D	E
1	**Capacitor Discharge**				
2	Enter or change the following details:				
3		Capacitor value	500	microfarads	
4		Discharging Resistor	100000	ohms	
5		Charging potential	10	volts	
6		Time steps	2	seconds	
7		Number of steps	30		
8					
9	Time	Charge	Potential	Current	Change in
10	s	microC	volts	mA	charge microC
11	0	5000	10	0.1	200
12	2	4800	9.6	0.096	192
13	4	4608	9.216	0.09216	184.32
14	6	4423.68	8.84736	0.0884736	176.9472
15	8	4246.7328	8.493466	0.084934656	169.869312
16	10	4076.863488	8.153727	0.08153727	163.0745395

A *spreadsheet* gives you the freedom to build a model on almost anything. You will have to be content with columns of numbers and graphs rather than nice graphics. You could build a spreadsheet to find the best volume to surface area for an animal. You could build a model to find the stopping distance of a car or to explore the dynamics of a population of foxes and rabbits. You can model the gas laws, the flow of heat through a metal bar or the lattice energies of compounds. Then again you can look at chemical equilibrium, radioactive decay, capacitor discharge or population genetics. You may need to create the spreadsheet models for yourself - but it's not *always* as hard as it sounds. The *Warwick Spreadsheet System* (Aberdare) is a package of ready made and (often too) clever spreadsheets, many of which are models. The *Essex Spreadsheets* (Essex) are well worth getting.

Ideas for using information technology in science

*U*sing IT *develops pupils' skills in handling information. These skills are not only valuable 'life skills' but they also enable them to delve deeper into science.*

Information technology itself consists of many tools. The ways in which they can help science are just as varied. It can help pupils to understand. It can offer a focus for discussion. It can provide access to rich materials. It may simply remove unnecessary effort - which on the surface saves time, but it also brings pupils closer to using what we call 'higher order skills'.

This section shows the very many points in the science curriculum where IT can help science teaching. I hope readers will be encouraged to build some of them into their teaching and get a measure of the value that IT can add to science.

Key to this section:

This is a topic heading ...

... which will be followed by a stimulus for a science activity.

... and followed by a **description of the hardware** or software that can be useful. Where appropriate, the description will give a reason why it is being used here.

Specific software titles are listed so: **Genetics** (Mac / PC - TAG). Which shows the machine availability as well as the supplier. Addresses and phone numbers are in the *Reference Section*.

*Then, there are the small print **footnote** entries showing:*

- *What **level** the activity will be suitable for:*
  Key stage 3 means ages 11-14, Key stage 4 ages 14-16 Advanced level ages 16-18.
- *What information technology skills are involved. This may be: Handling information, Communicating using IT, Measuring, Controlling and Modelling with IT.*
- *Where, if anywhere, to find more information.*

Animal identification keys

Build a key to identify a set of mammals, birds or reptiles.

You can use a **branching database** program to create an identification key. It can help pupils to structure their observations and it does so in an engaging sort of game. It's a good idea to set some constraints when they play this game. In this way you can get the pupils to focus on certain features of the creatures. For example you could pay special attention to their appearance - their skin, feathers or how many wings they have. You might instead focus on egg-laying, feeding or habitat. Such an exercise is a valuable observation activity at almost any level of work - even advanced level.

Key stage 3+. Handling information using a Database program See ...

EXAMPLE ONLY

Does it live in water?
Yes → Whale?
No → ?
Yes → Rat
No → Lizard

See 'Software for Teaching Science' for more detailed reviews of the titles in this section

Choosing software, like choosing any teaching book is subject to taste and teaching style. Where software has merit and has been used successfully I have recommended it. Otherwise the comments are negative or even neutral. For detailed reviews on current software for science, see the companion guide - 'Software for Teaching Science', details on page 127. Care has been taken to get all details correct, but things change, older titles get deleted, so please check with the suppliers. Many CD-ROMs available through education suppliers come with after-sales support. Details on page 127-129.

Using IT

Section

3

Using IT in ... everyday science

Revision tools and question banks

Publishers have spotted the potential of IT for exam preparation. Stanley Thornes' **Questions** (PC CD-ROM) and Doublestruck's **Exampro** (PC - Tel: Tel 0184 898969) help you to print customized exam papers. Both are worth seeing. **LETTS GCSE Science Revision** (PC CD-ROM for age 15+, from AVP) are the famous guides - gone weak on disc. **QM Web** (PC - QuestionMark) is a tool for creating Internet based questionnaires, tests, surveys and quizzes. For students there is **Exam Tutor** (All CD-ROM Granada), the **Abbey A level Revision Software** (PC - Trotman www.trotman.co.uk), **Acacia Revise** (PC CD-ROM - Dorling Kindersley), **Swift Test Software** for Biology A/GCSE levels (PC - Swift 0171 731 4108), **It's Biology** (PC CD-ROM A level - Tel 01572 822278) Don't even try to seek and ye shall find more.

Writing for science

Make a poster for the science corridor

Using a word processor in science is seriously underrated. It's an excellent way to get pupils to work together to plan, draft and refine their work. They do this at the computer much better than with pen and paper. As a starter exercise, the pupils might explain and illustrate how science affects everyone. They can focus on the advances in medicine, fabrics, food production, home entertainment or our use of computers. They can use a **graphics program** to prepare diagrams or a **scanner** to scan any useful pictures into the computer. The pupils can then place the diagrams on the page, print their work or even publish it on the **Internet**. Or use a **multimedia authoring program,** such as Hyperstudio (PC / Mac / Acorn - TAG), to present your information. Such programs allow you to use text, diagrams, sound and video.
Key stage 3-4 Communicating using word processing / DTP programs
Idea from Kaleidoscope (Heineman)

Plan an experiment

Use a word processor to prepare an experiment planning worksheet. The sheet can have questions such as 'what will you measure' and 'what do you expect to happen' which can guide the pupils through planning an experiment. Unlike worksheets you don't print this. Instead, the pupils use it on the screen. To get added-value, have the pupils working together.

> **Pupil Worksheet**
> See the
> **Word Processor** topic

Write up an experiment

Use a **word processor** to prepare a worksheet for writing up an experiment. The sheet will have questions such as 'what was the aim of the experiment', 'why do you think you got this result' and 'how could you have improved the investigation?' This sheet or template can prompt pupils to add a results table and draw a graph. If the pupils' IT skills are good, they will be able to include graphs they had prepared using say, a spreadsheet or data logging program.

Fill in the gaps exercises

If pupils have access to **word processors** there are opportunities to use them for cloze, gap filling and similar text exercises. Here you create the exercise, save it on disc and get the pupils to complete it on screen. Most underrated is the '**Developing tray'** or today's version called **WinTray** (PC from LETTS) or **Sherlock** (PC - primary look and feel from Topologica) This a gap-filling exercise in the computer, but specially written for the purpose. The program offers the pupils hints, reassurance and a scoring system - all of which seems to drive them well. As in most word processing work, collaborative work is strongly recommended.
Key stage 3-4 Communicating using word processing / multimedia programs

General purpose software for science

Choosing software

Word processor, database, graphics and spreadsheet programs are the most widely applicable tools for teachers' and pupils' science work. You might also have programs for data logging and control. For most uses of IT in science that's enough but still too many.

There is a way forward. Firstly, if you use modern computer systems you will find each program shares a common way of working such that, word processor or spreadsheet, you can progress from one to the other easily. Secondly if you use programs from the same manufacturer you'll find their different programs much more alike. Thirdly, if you use **integrated software**, i.e. a bundle containing say, a word processor, database and spreadsheet you'll find learning easier still - the links between the elements in the bundle tend to be transparent and therefore easy.

Using IT in ... everyday science

Where can I get ready-made diagrams?

Many graphics programs come with a library of pre-drawn pictures or **clip-art**. You may find pictures of animals and plants, human anatomy, apparatus, chemical structures and so on. Both teachers and pupils can use these pictures to illustrate their work. You would need a library of thousands to cope with every need, but with a little skill you might for example, turn a picture of a flask into one with tubes and a bung.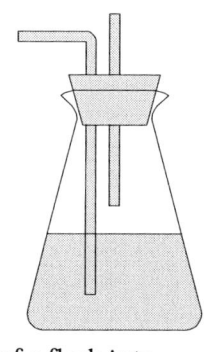

A rich catalogue of clip-art comes with professional graphics programs such as **CorelDraw, Designer** and **Arts&Letters** (PC Windows). More relevant is the **SSERC** library (Disc and CD-ROM PC / Mac / Arc - **SSERC**) which is a large and extremely useful collection of science apparatus diagrams. **Crocodile Chemistry** (RM) can even do experiments with its diagrams. You might take a look at **Graphics Explosion** (CD-ROM - TAG) or **Bitfolio** (CD-ROM PC / Mac / Arc - mail) **Sherston Clip Art Library** (PC / Arc - Sherston) is a collection for lower school groups. Multimedia programs and word processors often come with a collection of art. **Science Activities** (PC / Mac - TAG) are ready-made 'worksheets' for use with the ClarisWorks integrated package - a few of these are worth adapting to your way of working. There are some useful 'models' to experiment with though they demand very little from the student.

Computer graphics and scanners

If you prepare worksheets, posters and leaflets on the computer, at some point you'll feel a need to add graphics to your work. On the one hand, you may be skilled enough to draw your own pictures using a graphics program. If you have a picture on paper, why do it again? If you browse through the computer catalogues you'll find an accessory called a **scanner**. Both teachers and pupils should have access to one of these. It can take a picture from a page and the software can help you place it on the printed page. For little money you can get a hand held scanner which can easily cope with the need to put a small diagram on your page. There are several small and cheap page scanners which can 'read' the text of a document, produce a file you can edit and allow you to 'recycle' the ideas on it.

Find out what you can about ...

Having an encyclopaedia accessible from every science room in the school is, I think, a worthy aspiration. That is easily possible using IT but you have quite a choice: **Compton's Interactive Encyclopaedia** (CD-ROM PC - mail) provides informative articles and it is quite friendly while the **Grolier Encyclopaedia** (CD-ROM PC / Mac - TAG) is a bit hard. They pale beside **Information Finder: World Book Encyclopaedia** (CD-ROM PC) - which is easy, clear accessible and maybe the best. For users of Acorn computers, **Hutchinson's Science Encyclopaedia** (CD-ROM PC / Arc - TAG) has the distinction of being British. But see Microsoft's **Encarta** (CD-ROM PC / Mac - mail order) - a beautiful and all-subject electronic encyclopaedia. It throws up a surprising amount of science fact - even if it is hard to read. A separate resource pack is worth missing. Steck-Vaughn's **Science Encyclopedia** (PC CD-ROM on mail order from Andromeda) is a good choice for pupils aged 10-15. **Science Navigator** (Mac/PC for age 15+ from TAG) is a concise science encyclopaedia.

Eyewitness Encyclopaedia of Science 2 (CD-ROM PC / Mac - mail order) - a Dorling Kindersley title is still a bit shallow for an encyclopaedia. Finally, **The McGraw-Hill Encyclopaedia of Science and Technology** (CD-ROM PC - mail order) boasts 117,000 terms and definitions - useful but very expensive. **The New Scientist** (PC CD-ROM, age 16+ from SCET) has access to the magazine's full text and diagrams from April 89 to date. This is also available on-line, on subscription. It is not cheap.

Find out, share and communicate

The Internet lets us read about what scientists are doing, ask for advice from colleagues, and even publish our worksheets. It helps us to share information we normally share using meetings, memos, books, CD-ROM and phone calls. The reference section suggests a few places to see the things that are possible. You really need to visit a place for teachers and hop off from there: Find Internet links at www.rogerfrost.com and the ASE at www.ase.org.uk

Key stage 3-4 Handling information

Using IT in ... animal biology

Animals

How are animals different?

You can find ready-made **databases** for data search activities on almost any 'life' topic - be it birds, dinosaurs or minibeasts. You can explore how different species are adapted to their environment. You can find out which animals compete for the same foods. You might ask: which animals lay eggs, which suckle their young? Which live in water? Which live on land? Which animals share the same habitat?

Having access to computer-based data enables you to cut across the boundaries you find with data in books. You'll often finding new and interesting patterns. For example, is there a pattern between the number of legs that the animal has and its habitat?

You can start off by choosing an animal and finding out what you can about it. You might then search through a computer database to find those animals matching certain characteristics. Later you might look for patterns by plotting graphs - for example, you might plot an x-y graph of gestation period against the size of the animal. The **Key database files** (All machines - AVP) which are ready-made databases with masses of data on **Birds of Britain, Freshwater Fish, Seafish, Mammals and Minibeasts**.

> **Pupil Worksheet**
>
> See the Database topic

CD-ROM have video, photographs and sound. There are legions of these. **Creepy Crawlies** (CD-ROM Mac / Arc / PC - AVP) is a catalogue of creepy creatures for lower school projects. **Garden Wildlife** (Age 7-11, CD-ROM PC / Arc - Anglia) is better. It illustrates many garden creatures. **Bug Adventure** (CD-ROM PC - Guildsoft) is a home title about the world of insects. **British Birds** (CD-ROM Arc / PC - AVP) has photographs, sounds and film of all the British birds as well as curriculum materials. **Mammals: a Multimedia Encyclopaedia** (CD-ROM PC / Mac - mail) is a dated National Geographic Society disc detailing 200 animals. **Multimedia Animals Encyclopaedia** (CD-ROM PC / Mac - Education Interactive) contains pictures, and covers habitat, diet and a taxonomy of 2000 animals. **The Encyclopaedia of Mammalian Biology** (CD-ROM PC - McGraw-Hill) is a massive reference on mammals with text, pictures, data and film. It's a good resource but fiddly to use. **Dictionary of the Living World** (CD-ROM Mac / Arc / PC - AVP) is a quick reference tool, though it is brief on detail. **World Alive** (CD-ROM PC / Mac - Education Interactive) has lots of silent movies on a selection of animals. **The Zoo** (CD-ROM PC / Mac - Media Design Interactive) is about uninteresting happenings behind the scenes at the zoo. **Zoo Guides** (CD-ROM PC/Mac - mail) focuses on the mammals of Africa. **Animals of the World** and **Butterflies of the World** (CD-ROM PC - AVP) are possible library titles. **In the Company of Whales** and **Whales and Dolphins** (CD-ROM PC / Mac - mail) are for fans only. **Eyewitness Encyclopaedia of Nature 2** (CD-ROM PC / Mac - mail)

ought to have a search facility but it does the idea of food webs well. **Eyewitness Cat** and **Bird** (CD-ROM PC / Mac - mail) are good enough for the library, but again they have no search facilities. **Exploring Nature** (Age 9-13, CD-ROM PC / Arc - Hampshire) allows you to explore an environment equipped with a field book, notebook, map and sensors - above average. **Woodland** (CD-ROM Arc - TAG) is a picture book detailing British woodland life, but with poor search features. **Woodland Birds** (CD-ROM PC / Arc - AVP) is better. **Dangerous Creatures, Endangered Species, Dinosaurs** (CD-ROM PC / Mac - AVP) are three 'Microsoft Home' titles where the presentation is way above average. **Lost Animals** (PC CD-ROM for age 12+, Ransom) is a virtual museum with 50 extinct animals. **3D Dinosaur Adventure** (CD-ROM PC - Guildsoft) shows animated dinosaurs in gimmicky 3D and is mostly for fun. **Dinosaur Discovery** (CD-ROM Mac - Kimtec) catalogues 150 dinosaurs with very technical details and maps. **Survival – Mysteries of Nature** (Mac/PC/Acorn CD-ROM, age 8-14 from Anglia) is a fair, documentary-style title on the themes of flight, hunters, the senses as well as having animal data too. **How Animals Move** (Mac/PC CD-ROM for age 15+ from TAG) is well placed on a CD-ROM - it is quite hard though. **Wide World of Animals** (PC CD-ROM, age 11-16 on mail order) is a database of rare and other animals - well worth seeing.

Key stage 3 Handling information using a Database program

Animals: survey

What patterns are there in the features of animals?

As shown above, there's a case for using ready-made **databases**. The idea there is that the pupils get to analyse some data. On other occasions you will want pupils to learn how to collect and structure their own data.

Database table: animal survey						
Animal	**Type**	**Legs**	**Wings**	**Habitat**	**Babies**	**Skin**
Mouse	Mammal	4	0	Field	Live	Fur
Robin	Bird	2	2	Wood	Egg	Feathers
Spider	Arachnid	8	0			

First establish the questions you wish to answer as this determines the data you collect. You might record whether the animals live in water, how many legs they have or whether they are covered with feathers, scales or fur. After entering the data into a database you can easily sort the animals into groups. You can group them according to how many legs they have and then start to see what else they have in common.

Key stage 3 Handling information using a Database program

Using IT in ... animal biology

Animal identification keys

Build a key to identify a set of mammals, birds or reptiles.

You can use a **branching database** program to create an identification key. It can help pupils to structure their observations and it does so in an engaging sort of game. It's a good idea to set some constraints when they play this game. In this way you can get the pupils to focus on certain features of the creatures. For example you could pay special attention to their appearance - their skin, feathers or how many wings they have. You might instead focus on egg-laying, feeding or habitat. Such an exercise is a valuable observation activity at almost any level of work - even advanced level.

Does it live in water?

Yes / No

Whale? / ?

Yes / No

Rat / Lizard

See the Branching Database topic

Key stage 3+. Handling information using a Database program

Animal behaviour

Are animals more active during the day or during the night?

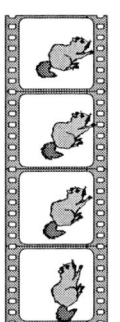

Use a sensor to monitor the activity of an animal in its nest. The computer can show you a line graph with periods of activity and non-activity. From this you can deduce the times when they are active. For example, if you monitor the light level using a **light sensor**, it should respond when the animal moves near it. You might also monitor the **infra-red** or **temperature** levels - as both provide clues as to when the animal is in its sleeping quarters.

Key stage 3 Measuring using sensors

Write an account of yourself as a bee, communicating with bees in a hive.

You might ask pupils to summarise their work on a topic by writing imaginatively and extensively. The **word processor** is not only useful for this but it is a valuable focus for collaborative work. The quality of pupil-talk in such situations is well worth experiencing and evaluating.

Bees

Bees are social insects

Key stage 3 Communicating using a Word processor
Idea from Kaleidoscope (Heineman)

Animals: birds feeding

How do birds distinguish their food?

Birds express a preference for certain food colours. Explore this idea by setting out two bags of different coloured nuts on a bird table and record the bird activity, at each bag, with the help of a sensor. Each bag of nuts can be attached to a **position sensor** such that each time a bird disturbs it you record an event on the screen. If you have only one such sensor you might arrange things such that the sensor is pulled one way or the other - depending on which bag is being visited. Using the sensor with a data logger will allow you to sample over a long period of time - longer than you would ever attempt without one.

Similarly you can use a position sensor (or a light gate) to simply count the number of birds arriving at a bird table. Remember to set up the data logger to count events, rather than record their magnitude.

Key stage 3 Measuring using sensors

Animals: keeping warm

Why do some animals huddle in cold weather?

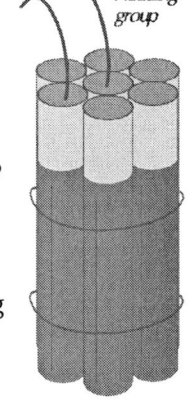

Huddling group

Use **temperature sensors** to measure the temperature changes in "keeping warm" experiments. You might investigate questions such as: why do penguins huddle? Or how much warmer does fur help keep animals? Does fur-up work better than fur-down? Does fur still work when it is wet? Using two temperature sensors you'll be able to take two sets of reading simultaneously. A graph is drawn while you are doing the experiment this provides a perfect picture of its progress. This graphical feedback is particularly helpful in investigative work - for example, there is a time saving which allows you to extend the work: so you might look at how animals keep cool - for example, how do elephants' ears help them to keep cool or even, which loses heat faster, a large or a small animal?

Key stage 3-4 Measuring using sensors

Using IT

Section

3

Using IT in ... animal biology

What is the 'best' ratio of volume to surface area for an animal?

We lose heat from our skin. The greater the surface of skin exposed to the cold air, the more heat we lose. Is there a particular size or shape of animal which will lose less heat? **Spreadsheet** programs easily handle large numbers of calculations and can help us to find the 'best' ratio of volume to surface area.

First assume that an animal is a cube and then enter a series of possible lengths, breaths and widths into the program. Calculate the volume and the surface area of each size and plot an x-y graph of the surface area against volume. From the graph you can find the shape or size which conserves heat best. Incidentally, you can work out the volume using V= lbd and work out the surface area using SA= 2lb+ 2bd.

	A	B	C	D	E
1	**Surface area to volume ratios**				
2	Animal height	Animal width	Animal length	Volume	Surface area
3	1	5	1	5	20
4	2	4	2	16	32
5	3	3	3	27	36
6	4	2	4	32	32
7	5	1	5	25	20
8	1	5	5	25	60
9	2	4	4	32	48
10	3	3	3	27	36
11	4	2	2	16	24
12	5	1	1	5	12
13					
14					

Key stage 3-4 Modelling using a Spreadsheet
Idea from Howard Flavell and Maurice Tebutt's 'Spreadsheets in Science' (John Murray)

Control - aquaria and fermenters

A computer controlled tropical aquarium

A topic on microelectronics is one place to look at the use of control systems. A topic on the similarities and differences between animal and computer **control systems** is another. For example, you might build a system to control the temperature in an aquarium. You first set up a temperature sensor in the aquarium and get a buzzer (an alarm) and a low current heater. Next, by writing a simple program in a control language, you can get the system to respond to a drop in temperature - it might for example, set off the buzzer and power the heater.

If you have computer sensors which detect pH and oxygen, for example, you can use them similarly. You might reasonably 'control' a fermentation process for a biotechnology project.

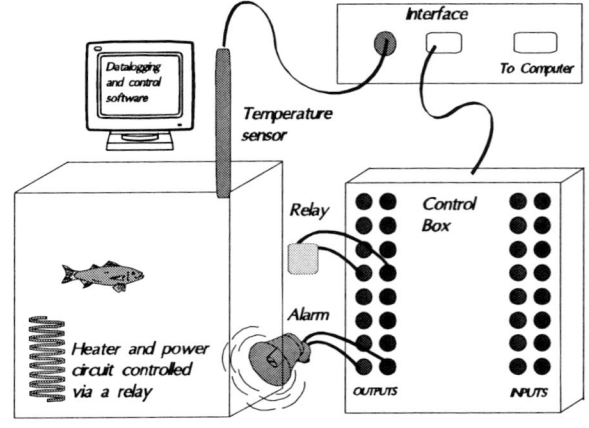

Key stage 3-4 Measure & Control using Control technology

Why is warm water bad news for tropical fish?

The amount of oxygen dissolved in a liquid decreases with increasing temperature. You can show this effect by using **sensors** to measure the **temperature** and **oxygen** level of water as it is warmed. By using the sensors you should obtain some useful data in an otherwise difficult investigation.

Key stage 4 Measuring using sensors

SOFTWARE and CDROM REVIEWS can be found in "SOFTWARE FOR TEACHING SCIENCE" © IT in Science

Using IT in ... human biology

Body organs

What's inside the human body?

There are many computer 'books' about the human body. **BodyMapper** (Age 7-13, Arc / Mac / PC - TAG) - an above average program where pupils can read about the body as well as compare their measurements and eye colours. **Bodyworks 6** (CD-ROM PC - AVP), **ADAM - the inside story**, **ADAM Essentials** (CD-ROM PC / Mac - AVP) are colourful and detailed anatomy atlases. They are worth having for the library - they need work to use them in class. There are many drawings you can copy and use and the plentiful information is often up to medical student standard. **The Living Body** (CD-ROM PC - AVP) is a dull offering despite its diagrams, animation, photographs and photo-micrographs. **Bodywise** (Age 10-14, CDROM Arc - Sherston) is a much better choice for school. **Understanding the body** (Age 11-14, CD-ROM PC / Mac / Arc - Anglia) is also better suited to the age group. There are pictures and text you can use. **The Ultimate Human Body 2** (CD-ROM PC / Mac - AVP) is well put together. It allows you to explore the body systems and is excellent for library work. **How your body works** (CD-ROM PC - mail) is a 'home' title but with some brilliant 3D animation. **The Human Body** (CD-ROM PC - AVP) is worth a look too. **3D Body Adventure** (CD-ROM PC - Guildsoft) is mostly fun: you watch body movies wearing 3D glasses - incredible and useful to demo though. **Interactive Skeleton** (PC CD-ROM from SCET) and **Dorling Kindersley's 3D Skeleton** (PC CD-ROM from mail order) lets you take a bone and spin it round on the screen. Nice idea but there is not much else to engage interest. **Magic Bus explores the Human Body** (CD-ROM PC / Mac - AVP) is easy, and *too* much fun.

Key stage 3-4, A level. Modelling using Graphics and Simulation programs

Make a key to identify the human organs.

> See the Branching Database topic

Start by sorting out some models or pictures of body organs, thinking about where they are found and what they do. Next, use a **branching database** program to create a key to help others identify the organs. A branching database can be the centre of an engaging and valuable observation exercise at almost any level of work - even advanced level. Test the key on another group.

Key stage 3-4 Handling information
Idea from Information Technology in Science (MEU Cymru)

A human body cut-out.

> Pupil Worksheet
> See the Graphics tools topic

There is a computer equivalent to those exercises where pupils cut out pictures, colour them in and then put them in their correct places. Cut-outs seem to take an inordinate amount of pupil time. However, by using the clip-art that comes with many **graphics programs** you can create your own on-screen cut-out exercise. You will often find, as shown elsewhere in this book, ready-made pictures of the various body organs which you scatter around the screen. The pupils have to place the organs in their correct places and label them. The example shown is a touch more interesting - the pupils also have to re-scale the organs to their correct sizes and boxes on the screen help them do this correctly. When it comes to colouring in, whatever you make of that educationally, the computer can do that very quickly. **Skeleton** (Acorn from AVP. PC version is dated however) helps you create a life size human skeleton - nice activity this.

Breathing

How does exercise affect your breathing?

Use a **pressure sensor** and **stethograph** (or a position sensor and spirometer) to monitor a pupil's breathing before and after exercise. The computer display can show you not just the rate and depth of breathing but also the pattern of inspiration and expiration at rest and during oxygen debt.

How strong are your lungs?

Use a **pressure sensor** to take measurements of lung pressure during normal and forced expiration. Try to relate the measurements to the height or chest size of the individual. You can take these measurements and build up a database of the whole class. See the section on **Genetics and Variation** later, for a substantial class project idea.

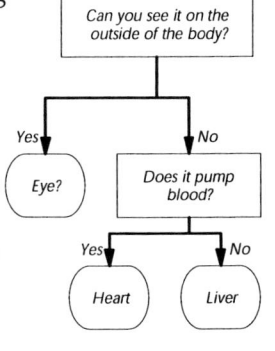

Using IT in ... human biology

Do inhaled and exhaled air have the same amount of oxygen and water?

Using computer sensors allows us to show, with a graphic display, the difference between inhaled and exhaled air. You can gain clues as to the amount of oxygen removed from the air. You simply use an **oxygen sensor** to measure the oxygen level as you breathe and re-breathe the air in a plastic bag. (Care. Don't fall over!) You can then flush the bag with another gas and watch the oxygen level drop - showing that we remove comparatively little oxygen from the air.

You can also see the change in air humidity during breathing. For this you use a **humidity sensor** in an otherwise similar experiment.

Key stage 3-4 Measuring using sensors

Exercise and circulation

How does exercise affect the pulse rate?

Use a **pulse sensor** to measure the pulse. You'll be able to see, for example, how long it takes for the pulse to recover after exercise. The activity can be highly recommended.
Using an **ECG sensor** you can gain some further insight into the heart.

You can also tape **temperature sensors** to different parts of the body and see how their temperature changes during and after exercise. Or tape the temperature probes to a pair of muscles and see how the temperatures change a) during an oscillating exercise b) during a static exercise.

Key stage 3-4 Measuring using sensors

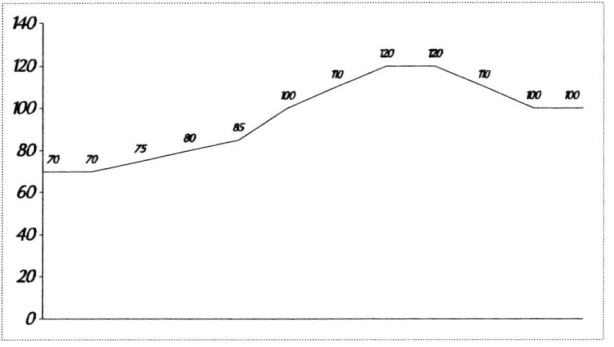

How quickly does the pulse return to normal?

Take pulse readings of individuals in the class at rest, just after exercise and then at four and eight minutes later.
Use a **spreadsheet** program to record the results. The program can show the class average very easily. It can also plot everyone's results on a graph and for example, show how the recovery of the pulse varies across the class.

	A	B	C	D	E
1	**How fast do our pulse rates recover?**				
2	**Name**	**Resting**	**After exercise**	**After 2 minutes**	**After 4 minutes**
3	Mandy	85	130	90	84
4	Andy	83	135	88	83
5	Sandy	80	120	85	81

You might calculate a fitness index for each member of the group based on the Harvard step test, or whatever test is in vogue. Briefly, this consists of doing five minutes of step-ups followed by four minutes of resting. During the rest you record the pulse count over three minutes. Use the calculation here and see how close you are to the average fitness index of 65.

	A	B	C	D	E	F	G
1	**How fit are you?**	Do five minutes of step-ups and record your pulse					
2	**Name**	**Exercise time x 100**	**Pulse from 1 to 2 mins**	**Pulse from 2 to 3 mins**	**Pulse from 3 to 4 mins**	**Sum of pulse counts**	**Fitness index**
3	Mandy	36000 secs	Index =	Duration of exercise in seconds x 100			
4	Andy						
5	Sandy			Sum of pulse counts			
6							

Key stage 3-4 Handling information Spreadsheet

SOFTWARE and CDROM REVIEWS can be found in "SOFTWARE FOR TEACHING SCIENCE" © IT in Science -

Using IT in ... human biology

Cardiovascular System

The workings of the cardiovascular system is the topic of the simulation program, **Cardiovascular System** (Mac/PC Windows - TAG).

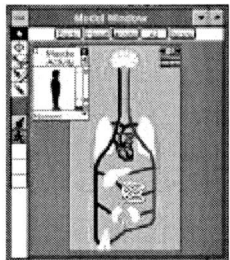

With this you can see the heart at rest and during exercise. You can also explore in detail, the anatomy and workings of the CVS. A bonus feature is that you can connect yourself to the computer and use your personal pulse rate to drive the display on the screen. Must be seen - ask about the later range - *Gateways*.

Key stage 4 to A level Modelling using a Simulation program

Energy requirements

How much energy do you use in a day?

Find one of those data tables which show how much energy different activities require. Make a complete record of everything you do in a day and how long you spend doing it. Enter all this, as a table, into a **spreadsheet**. Using this program, you can calculate your daily energy requirement: you type a formula in one column of the spreadsheet to calculate how much energy each activity uses. You then total this column to find your energy requirement.

	A	B	C	D
1	**Energy requirements**			
2	Activity	Energy use kJ per hour	Duration hours	Energy total kJ
3	Running	1600	0.5	800
4	Cycling	1250		
5	Walk upstairs	1000		
6	Walking	800		
7	Light activity	600		
8	Studying	400		
9	Sitting	300		
10	Sleeping	200		
11	TOTALS			800

You can use the same spreadsheet to find out how much energy you would use if you a) took part in a bicycle race day. b) spent the day running a Marathon. c) spent the day watching the Marathon on television.

For ready-made programs where you enter the kinds of exercise you take, the type of foods you eat and find your energy balance, see the section on Food: nutritional value.

Key stage 4 Modelling using a Spreadsheet
Idea from Blackwell Modular Science

Eyes: testing sight

How much do our eyes vary?

Do eye tests on a broad sample of individuals. Enter the data into a **spreadsheet** table and plot a bar chart to show the spread of the results. At the same time you can record possibly-related factors such as age, gender and hair colour.

	A	B	C	D	E	F
1	**Are there any patterns in how well we see?**					
2	Name	Eye test 1	Eye test 2	Hair colour	Age	Other
3	Mandy					
4	Andy					
5	Sandy					

You might then try to find a correlation between say, age and eye-sight. For example, you might plot an x-y graph of age against eye test result.

Key stage 3 Handling data using a Spreadsheet

Growth / Making Babies

When do we grow fastest?

Science work is rich in exercises on drawing and asking questions about existing data. As an example, "use the data provided to plot a graph of the growth of a boy/girl from birth to 20 years. At what ages do they grow fastest? What can you say about the growth rate when the person reaches 20 years old?" By entering the table directly into a **spreadsheet** we can quickly get to the point of a fairly onerous exercise. Using the program, it is very easy to produce an x-y graph of growth against age. It is also a fairly easy matter to calculate and plot the **rate** of growth in each year of life.

	A	B	C	D	E
1	**How fast do we grow?**				
2	Age	Boy - mass	Girl - mass	Growth rate kg/year	Growth rate kg/year
3	0	3	3	0	0
4	0.25				
5	0.5				
6	0.75				
7	1				

For sex education, **The Facts of Life,** (CD-ROM PC - TAG) is all you ever needed to teach about sex - students can read about the facts, sexuality, religion, parenting and relationships. **The Baby File** (CD-ROM Mac - TAG) is a narrated slide show on gestation. There is also **The First Year of Life** and **Years One to Three** (CD-ROM Mac - mail) and **Nine month miracle** (CD-ROM PC / Mac - mail).**Growing Up Together 2** (Acorn/PC for age 8-11 from Granada) is for personal & social education.

Key stage 3-4 Handling data using a spreadsheet. Growth idea: 'Active Science'.

Using IT

Section

3

Using IT in ... human biology

Heart and the blood

Make a poster about the things that can go wrong with the heart and circulation.

Use a **word processor** or **graphics program** to prepare the poster. To add a diagram, you can use the clip-art from a graphics program or scan-in some images from paper. The poster might also encourage people to improve their health. This sort of activity is best done in groups.

Key stage 3 Communicating using word processing / DTP programs
Idea from Kaleidoscope (Heineman)

You are a red blood cell and you've just seen an advertisement for a job as Oxygen Transporter...

To summarise their work on say, blood, the pupils might be asked to write an application for this job. They might list the qualities that make them suited to the work. They can use a **word processor** to write their job applications and, unlike work with pen and paper, they can collaborate to develop their first draft into a more refined job application.

Key stage 3-4 Communicating using a Word processor
Idea from Bath Science (Nelson)

Homeostasis

Build a model to simulate a life support unit

People trust their lives to computer controlled machinery such as kidney machines and life support units. Using **control technology** you can build a simple model of one and help to reinforce ideas about homeostasis. Your model might pump air when the oxygen level drops or it might pump alkali ('bicarbonate') when the pH drops. You need a control box, a sensor to measure oxygen or pH and a device (a pump of sorts) to create this intriguing model. You can use the same idea to run a biofermenter in a biotechnology project.

Key stage 3-4 Measure & Control using Control technology

Health

Do a survey on peoples ideas about health.

Personal health, awareness of the causes of various illnesses and the effect of diet and exercise on the heart are good topics for a survey. You can use a **word processor** to draft, test and finalise a questionnaire on health. It is worth taking time to get this right before going on to collect the data as success in this kind of project really depends on it. Too often pupils fail to identify the key questions that need an answer or they collect data without an agreed set of headings or an agreed format. For example, it's often better to record answers to questions as a score, as numbers are easier to graph than words.

Finally, you can enter the survey results into a **database program**. The database program allows you to analyse the data by sorting and graphing.

	A	B	C	D	E	F	G	H
1	Health survey							
2	Person	Smoke per day	Drinks per week	Cause of bronchitis	Cause of stomach ulcers	Cause of heart attacks	Age	Gender
3	A	0	1	Flu	Alcohol	Fried food	15	M
4	B	10	0	Smoking	Vinegar	Eggs	21	M
5	C	0	20	Smoking	Stress	Fatty diet	15	F

For example, if you had recorded people's ages and ideas about the causes of heart attacks you could look at the results to find out if ignorance was age related.

Key stage 3-4 Handling information using a Database program
Idea from Folens Copymasters

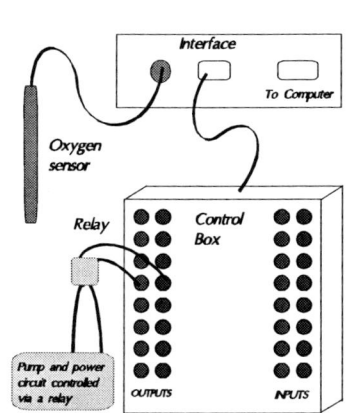

SOFTWARE and CDROM REVIEWS can be found in "SOFTWARE FOR TEACHING SCIENCE" © IT in Science

Using IT in ... food biology

Make a 'stay healthy' poster for a doctor's surgery.

Use a **word processor** or **graphics program** to prepare a poster for a doctor's surgery. The poster might include a useful plan of action for health. Use clip-art from a graphics program or images from a scanner to add any diagrams you need. Use the Internet to help your research.

Key stage 3 Communicating using WP / DTP programs

Why do they put fluoride in toothpaste?

Science work is rich in exercises involving drawing and interpreting graphs. For example: "look at the data showing the effects of adding different amounts of fluoride to the water supply. What does it tell you about the effect of fluoride on tooth decay?" A **spreadsheet** can quickly draw a graph of this data.

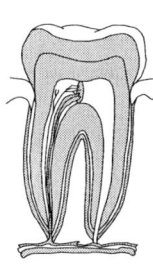

	A	B	C
1	**Percentage of children with this number of bad teeth**		
2	No. of bad teeth	High fluoride area	Low fluoride area
3	0		
4	"1-5"		
5	"6-10"		
6	"11-15"		
7	"16-20"		

In this case, you would plot an x-y graph of the amount of fluoride added against the number of fillings needed.

Key stage 3 Handling information using a Spreadsheet
Idea from Bath Science (Nelson)

Meiosis

For crystal clear animation of **Meiosis and Mitosis** see New Media's title of the same name. You can stop it running and talk about each stage.

Key stage 4 Modelling using a Simulation program

Nerves

Nerve (Arc - AVP) is a program where you can explore a simulation of resting and action potentials, the frog sciatic nerve and also plot the results on a graph. **Simnerv** (CD-ROM PC / Mac - ISBN 3 13 799604 X) does all this and more, right up to post graduate level and very capably too. Otherwise, many CDROM titles about the body cover the nervous system.

A level Modelling using a Simulation program

Food: breakfast cereals

How do breakfast cereals compare?

Pupil Worksheet

See the Spreadsheet topic

Doing a survey of cereals is a useful activity when learning about nutrition - you just need the data on some cereal packets and a computer **spreadsheet**. You might start by asking: which cereals are best for slimmers? Which cereals provide the most energy for your money? Which cereals provide the most protein for your money?

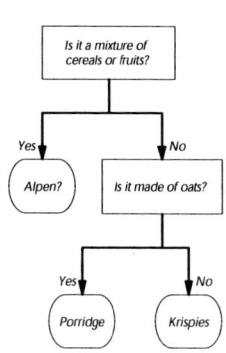

You type the cereal data into the spreadsheet. You can then draw a pie graph to show the make-up of a given cereal or a bar graph to compare all the cereals with each other.

	A	B	C	D	E	F	G	H
1	**Which cereal is best for energy?**							
2	Food	Main grain	Size	Price	Energy	Protein	Cost	Energy
3			g	p	kJ/100g	g/100g	/serving	/serving
4	Sugar Puffs	Wheat	450	115	1554	6	7.67	466.2
5	Strawberry crunch	Oats						
6	Porridge	Oats						

With little more effort, you can do the calculations to show those cereals which give the best value for money. As an alternative you might compare different brands of yoghurt or diary spreads.

Key stage 3-4 Handling information using a Spreadsheet

Make a key to identify different breakfast cereals.

A **branching database** helps you to structure information and build an identification key. The program can be the focus for an engaging observation exercise. You start by making a collection of cereals and sorting them into sets. You use the database by identifying questions which help distinguish the cereals. Questions such as 'does it contain wheat' are perhaps more scientific than 'does it have red bits' - so you should perhaps set some rules for the types of questions that are allowed.

Key stage 3 Handling information using a Database program
Idea from Data Handling in Primary Science (NCET)

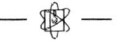

Using IT in ... food biology

Food: nutritional value

Do you eat the right foods?

Do you have a diet which is balanced in terms of energy, vitamins, protein and fat? **Diet analysis programs** can give you the answer with ease. These programs know the nutritional content of foods as well as the recommended daily amounts for each type of person. The programs can track the nutritional elements in the diet and produce graphs of various kinds to see if you're in balance. The hard part is keeping a note of everything you eat over a 24 hour period. You also need to estimate how much you eat - although the programs often suggest typical portions. Examples of these programs include **Diet Guide** (Age 9-13, Arc - Hampshire) which are favourites. **The Food Program** (PC - LETTS) is a strong offering while it's worth finding the very capable Nutrition section in the **Encarta** encyclopaedia (CD-ROM PC / Mac - mail). Other titles include **Food** (PC - AVP), and **Food Feedback** (PC - National Dairy Council) but the later can only do a rough analysis of your diet and misses the opportunity to do the sort of analysis you need in school. For advanced level see **Balanced Diet** (PC DOS - TAG) which is ugly and not even good. Latest and best is New Media's **Diet Analyser**.

It is perfectly possible to do the same job using a **spreadsheet** program. You will need to accept the limitations of this method: it is fairly hard work collecting the data, though it does work.

Key stage 3-4 Modelling using a Simulation program

	A	B	C	D	E
1	How much energy do I get from my food?				
2	Food	Energy	Typical portion	What I ate	Energy
3		kJ/100g	g	p	/serving
4	Carrots	1554	450	115	466.2
5	Tomatoes				
6	Beef				
7	Bread				

Food: chemical energy

Which food provides the most energy?

And which food provides the most energy for your money? To find out do a survey of different foods. Collect the energy information from 20 or more food labels. Also collect package sizes and prices. Your data is easily recorded by entering it into a **spreadsheet** program:

	A	B	C	D	E	F
1	Which food is best for energy?					
2	Food	Size	Price	Energy	Cost	Energy
3		g	p	kJ/100g	/serving	/serving
4	Yoghurt	450	115	1554	7.67	466.2
5	Rice Krispies					
6	Butter					
7	Margarine					

You might sort the table on say, energy content and plot a bar graph to compare the foods. You might also calculate the cost per 100g and so determine the 'best' value food for money.

Key stage 4 Modelling using a Spreadsheet
Idea from: Blackwell Modular Science / Science Scene(Hodder)

Measure and compare the energy content of foods.

We can burn samples of food to see how much energy they contain. We quantify this by burning a known amount of food and using it to heat a known amount of water. In this 'classic' science experiment, a **temperature sensor** can record the temperature change of the water as it is heated. Using a sensor allows you to concentrate on the techniques of controlling variables and preventing heat loss. Incidentally, should the food extinguish, the graph will provide a record of the loss in temperature - so you can still find the energy gained by the water.

> **Pupil Worksheet**
>
> **See the Spreadsheet topic**

If you wish, you can then enter the results into a **spreadsheet** program. Here you can calculate the energy obtained per sample of food. You can also draw a bar chart to compare the foods.

	A	B	C	D	E	F
1	Measuring the energy content of food					
2		Unit	Example	Peanut	Crisp	Pea
3	Mass of food	g	1.6			
4	Mass of food after burning	g	0.5			
5	Mass of food burned	g	1.1			
6	Amount of water heated	cm3	10			
7	Starting temperature of water	deg C	20			
8	Final temperature of water	deg C	31			
9	Temperature rise	deg C	11			
10						
11	Energy content of food	kJ/100g	508.2			

Key stage 3 Measuring using sensors

Using IT in ... food biology

Which bean gives the most energy for a given weight?

Suppose you were on an expedition and needed to take a minimal supply of beans as food. Which bean should you pack? To find out, you burn different beans and measure the temperature rise of a known volume of water. You then type the results into a **spreadsheet** program. You use the program to calculate the energy obtained per gram of bean. It's a simple further step to draw a bar graph of the energy released for each type of bean.

If you wish, use a **word processor** to write a report and put your results table and graphs into it.

	A	B	C	D	E	F
1	Comparing beans for energy					
2	Bean	Mass of bean	Temp of water	Temp after heating	Temp rise	Energy / gram bean
3	Broan					
4	Red					
5	Haricot					

Key stage 3-4 Modelling using a Spreadsheet

Food: value for money

Which is the 'best brand' of pop-corn?

When you pop corn not all the corn pops. Some brands of corn may yield more useful pop-corn and be better value for money. One way to find the best brand of pop-corn is to test them, weighing them before and after popping. You should also record their prices and pack size. All the results are easily entered into a **spreadsheet** where you use formulae to calculate the cost per 100g of pop-corn and the cost per 100g of good, 'poppable' corn. A bar graph comparing the pop-corns side by side can then be drawn.

	A	B	C	D	E	F
1	Pop Corn					
2	Brand	Weight of pack	Cost of pack	Weight of sample	Weight of popped corn	Cost of popped corn
3		g	£	g	g	£
4	Big Chief					
5	Amora					
6						

Key stage 3-4 Handling information using a Spreadsheet

Which is the 'best' value banana?

To find the 'best' value banana you need to measure how much of a banana is edible flesh. Weigh the amount of skin and edible flesh on different bananas and enter the results into a **spreadsheet** program. Use the program to calculate the amounts of edible banana by subtraction and then to work out the cost of the edible banana.

It's then quite easy to draw a bar graph comparing the different bananas. You may also want to 'measure' the quality of the bananas.

	A	B	C	D	E	F
1	Bananas					
2	Brand	Weight	Cost	Whole weight	Peeled weight	Cost of banana
3		g	£	g	g	£
4	Canaries					
5	M&S					

Key stage 3-4 Handling information using a Spreadsheet

Are large eggs better value?

To find the 'best' value egg you need to measure the amount of shell, yolk and white in different sizes of egg. The results can be analysed using a **spreadsheet** program. Use it to calculate the amounts of yolk and white by subtraction and then to calculate the cost of the edible egg.

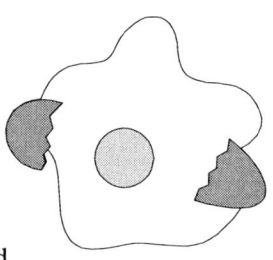

	A	B	C	D	E	F
1	Egg survey					
2	Type / Size	Weight of egg	Weight of contents	Weight of yolk	Weight of white	Cost per gram yolk
3		g	g	g	g	£
4	1					
5	2					
6	6					

You can then draw a bar graph to show the costs and relative amounts of yolk and white of different egg sizes.

Key stage 3-4 Handling information using a Spreadsheet

Food: cooking

How long does it take for egg white to harden?

When egg white hardens a change occurs which can be measured colorimetrically. You can use a **light sensor** in place of a colorimeter and monitor the progress of this change. If you were hardening the egg with heat, if would be interesting to measure the temperature at the same time.

Key stage 3-4 Measuring using sensors. See School Science Review March 89

Using IT in ... food biology

How long does it take for food to cook?

You can see how well heat penetrates food using **temperature sensors**. You place temperature probes into large and small potatoes and boil them in a container. You might then annotate the resulting graph to show the times when the potatoes are deemed to be cooked.

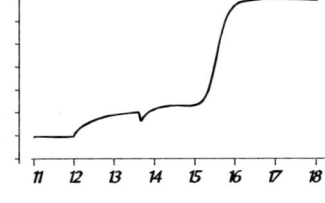

Pupil Worksheet
See the Computer Sensors topic

See the worksheet for another food example.
Key stage 3-4 Measuring using sensors

Enzymes and food

What affects the rate of enzyme-catalysed reactions?

You can use sensors to study the effect of trypsin on milk or lipase on fat. These two reactions involve turbidity changes and can be monitored using a **light sensor**. You can also study the effect of urease on urea, the making of yoghurt or the souring of wine. These reactions involve pH changes which can be monitored with a pH sensor. See the book **Data logging and Control** for details.

Key stage 4 Measuring using sensors

Make a poster on how cheese is made.

You may want to do this as a flow diagram, perhaps showing at what points in the process enzymes are active, and at what points microbes are active. A **word processor** or **graphics program** will help pupils to prepare their poster. This is best done as a group exercise.

Key stage 3 Communicating using WP / DTP programs
Idea from Kaleidoscope (Heineman)

Water

How do foods vary in their water content?

Measure the dry and wet weights of different foods. Then enter this data, as a results table, into a **spreadsheet**. The spreadsheet will help you calculate the water content of the food and produce a graph to compare them.

	A	B	C	D	E	F	G
1	Food water survey						
2	Food	Weight of dish	Dish + wet food	Dish + dry food	Wet weight	Dry weight	% Water
3		g	g	g	g	£	
4	Apple						
5	Cake						
6	Biscuit						
7	Meat						

Key stage 3-4 Handling information using a Spreadsheet

How do living things get food?

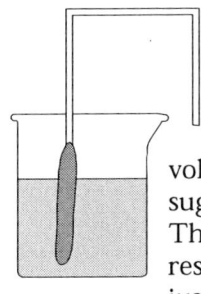

Do an experiment to illustrate osmosis. Use an **electronic manometer** or **pressure sensor** to monitor osmosis. The sensors can easily magnify the tiny volume changes in experiments using sugar solutions and Visking tubing. The sensors are sensitive enough to respond to changes occurring over just a few minutes.

Key stage 3-4 Measuring using sensors

Using IT in ... plant biology

Leaves

Which leaf shape gives the best ratio of surface area to volume?

Leaves respire through their surfaces. They can do this better if they have a large ratio of surface area to volume. To find out which leaf shape is best you need to represent, in a table, a number of leaves with different lengths, breadths and thicknesses. Working out the ideal leaf shape is easily done using a **spreadsheet** program. This can use your data to calculate the volume and the surface area of each set of sizes and it can also plot an x-y graph of the surface area against volume. From the graph you should be able to find the 'best' ratio of surface area to shape. Incidentally, you can work out the volume using $V = lbd$ and work out the surface area using $SA = 2lb + 2bd$.

	A	B	C	D	E
1	The best size for a leaf				
2	Leaf thickness	Leaf width	Leaf length	Volume	Surface area
3	1	5	1	5	20
4	2	4	2	16	32
5	3	3	3	27	36
6	4	2	4	32	32

Key stage 3-4 Modelling using a Spreadsheet
Idea from Howard Flavell and Maurice Tebutt's book of spreadsheets

Build a key to identify different leaves.

Collect a set of leaves from different trees - or just use some pictures on card. Sort the leaves into  types - paying attention to say, how many leaves are on a stalk, how many lobes the leaf has and whether the leaf is serrated or prickly. Next, use a **branching database** program to create a key to identify the leaves. The program helps you to structure a series of questions about the leaves and to build up a key. It's quite surprising how easily pupils get involved with this exercise - so do give it a try. When the key is finished, see if other members of the class can use the key. They may be able to improve upon it and remove ambiguities.

Key stage 3-4 Handling information using a Branching Database program

Plants: growth

Investigating plants

A particularly special resource is Homerton's **Investigating Plant Science** (PC CD-ROM - mail). It provides a framework where pupils are encouraged to design and carry out experiments  on plants. They have to pay due regard to the myriad of variables, the size of their sample, what they will measure and how they will display their results. There is also a bank of information with abstracts of past work. A deep and scientific approach to investigating plants.

Botanical Gardens (Mac - TAG) is a problem solving program on plant growth. You can experiment with, or model, the environmental conditions, record the results on a graph and form an hypothesis about the conditions required for plant growth. There is also section where you can design seeds. This a straightforward modelling program which is easy to recommend for lower school work. **Science Explorer II** (PC/Mac - for KS2 Granada Learning) has a plant growth simulation.

Plantwise (Arc - Sherston) is a surprisingly versatile program dealing with many aspects of plant structure and function for pupils aged from 10-15.

Key stage 3-4 Communicating using a Spreadsheet / Modelling using a simulation program

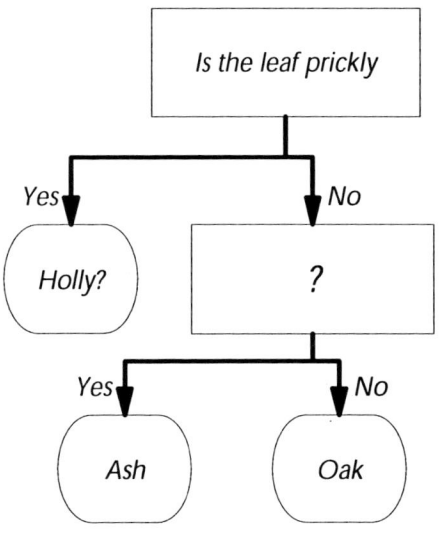

Using IT in ... plant biology

Does water always help the growth of a plant?

Compare the growth of similar plants. Give each plant a different amount of water and measure its growth daily. Record the results in a **spreadsheet** table.

	A	B	C	D
1	**Growing plants**			
2	Date / Plant	A	B	C
3	6.May			
4	7.May			

Use the program to prepare a bar graph to compare the plants. Draw an x-y graph to show how watering affects plant growth.

How fast does water travel during osmosis?

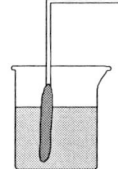

You can measure the rate of osmosis by recording change in volume with a **pressure sensor**. The sensor is •often sensitive enough to produce useful results - often in less than an hour.

Key stage 4 Measuring using sensors

Plants: light

How does the rate of photosynthesis change during the day and night?

Use an **oxygen sensor** to measure the oxygen content of either an aquarium or some water containing pond weed. Use a **light sensor** to monitor the light level and leave this running over say, a weekend. By using the sensors you should obtain some convincing evidence of the effect of light level on the rate of photosynthesis. Using similar apparatus and some coloured filters you can see how the rate of photosynthesis is affected by lights of different colour.

Key stage 4 Measuring using sensors

Make a poster to explain what happens in photosynthesis.

Use a **word processor** or **graphics program** to prepare the poster. You can use clip-art from a graphics program or images from a scanner to add any diagrams you need. This activity is best done working in twos or threes.

Key stage 3 Communicating using WP / DTP programs
Idea from Kaleidoscope (Heineman)

Simulating photosynthesis

Photosynthesis (Mac /PC - TAG) is a thorough, graphic simulation of photosynthesis. In this program you can explore the effect of variables such as light intensity, wavelength, humidity and temperature in a way which is difficult to do given the usual time constraints.

A level Modelling using a Simulation program

How does light affect how a plant grows?

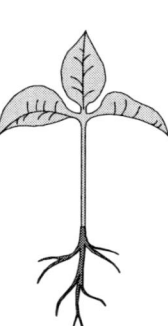

Use a **position sensor** with its lever arm tied to a plant, to study the growth of a plant over a few days. You can get a measure of phototropism by re-orienting the plant and seeing how it responds. If you can set up a couple of plants in this way, you can also study the effect of light and dark on plant growth.

Key stage 4 Measuring using sensors

Greenhouse temperatures

Build a system to control the temperature of a greenhouse.

A topic on microelectronics is the usual place to consider **control systems**. On the other hand you could make use of control in a biological context - after all there's a clear parallel with biological temperature control systems.

For example, you might build a system to control the temperature of a greenhouse. Set up a temperature sensor in the 'greenhouse' and collect together a buzzer (an alarm), a heater (a lamp) and/or a motor controlled ventilation window. Next, by writing a simple program in a control language, the system can be made to respond to a change in temperature. If you monitor the temperature continuously, you'll be able to judge how well the temperature is being controlled. It's interesting to see whether it's better to have a heating device, a cooling device or both.

Key stage 3-4 Measure & Control using Control technology

Arts & Letter

Using IT in ... plant biology

Plants and variation

How are plants different?

You can find ready-made **databases** for data search activities on plants. Having access to computer-based data enables you to cut across the boundaries you find with data in book form - finding interesting patterns.

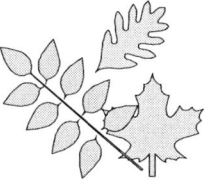

For starters, see the mass of plant data in the **Key** database file (All machines - AVP) on **Flowers and Vegetables**. The information here is useful but you will have to work out how to teach with it.

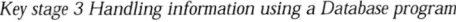

Key stage 3 Handling information using a Database program

Seeds and micro-organisms

Do seeds release energy as they germinate?

 Allow seeds to germinate in a vacuum flask and use a **temperature sensor** to measure the temperature change. See the book **Data logging & Control** for the details of this experiment.

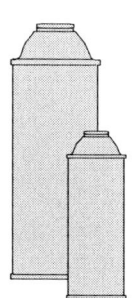

Key stage 3-4 Measuring using sensors

How fast do bacteria grow?

Use an **oxygen sensor** to monitor the growth of Bacillus subtilis over time.

Key stage 4 Measuring using sensors. School Science Review June 93

Does yeast release energy as it respires?

Place yeast in a vacuum flask and use a **temperature sensor** to measure the temperature change. As a control, monitor a similar flask containing killed yeast.

Key stage 4 Measuring using sensors

Are grass cuttings still alive?

You can monitor the **temperature** of grass cuttings using sensors. You collect the sample and put it in a vacuum flask to measure the change in temperature over time.
You might instead see how the temperature changes in a compost heap, haystack, beehive or ant-hill. A data logger will allow you to measure the changes over a long period of time.

Key stage 3-4 Measuring using sensors

Using IT in ... genetics and variation studies

Evolution / Genetics

What happens if you cross a ...

Computer programs can 'do' genetics experiments and help explore possibilities which are quite difficult to explore in real life. Those mentioned here provide more of an open-ended environment than is usual. **Blind Watchmaker** (Mac / PC - Capedia) is for advanced level students. It demonstrates the way living things can evolve - based on the famous book of the same name. **Drosophila Genetics / Pea Plant Genetics** (Mac/PC - Newbyte) are also for advanced level work. It is a very detailed computer model covering Mendelian dominance, phenotype ratios and more. It allows you to try your own genetics experiments and see the results in a shorter time-scale than you could normally.

Key stage 3-4 Modelling with a Simulation program

What makes the 'ideal' pig?

A group did a project to ascertain the factors required for breeding the 'ideal' pig. They went on to build a computer 'expert system' that could do this - a sort of flowchart which helps you find a solution to a question. They used the modelling program **Expert Builder** (PC Windows - AU).

Key stage 3-4 Modelling. In Information Technology in Science (MEU Cymru)

Plant strains

Are plants better adapted to certain environments?

In Biology work you might need to compare the growth of seedlings - either from an actual experiment or as a paper exercise from a book. For example, I found the results of an experiment comparing the growth of outdoor and standard varieties of tomato seedlings. Here, 25 seeds of each kind had been planted in two trays and allowed to grow. After a few weeks their heights were measured. It was a simple matter to enter the results into a **spreadsheet** and then use it to plot a histogram showing how the plants fared.

	A	B	C
1	**Growing seedlings**		
2	**Height of seedlings**	**Indoor variety**	**Outdoor variety**
3	"0-9mm"		
4	"10-19mm"		
5	"20-29mm"		
6	"30-39mm"		
7	"40-49mm"		

Key stage 3-4 Handling information using a Spreadsheet. Example from Folen's Copymasters series

Reaction time

What affects our reaction time?

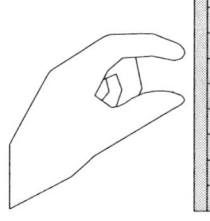

You might wonder whether if there is any relationship between the reaction time and factors such as age or fitness. Measure each person's reaction time, for example, by getting then to catch a ruler as it falls and then reading off the distance the ruler fell. At the same time, record the person's age, whether they wear glasses, whether they are good at sport, the time of day or even whether they have recently taken exercise. A **spreadsheet** can help record and analyse the results. Once you've typed the data into a spreadsheet, you can sort it on the basis of age or reaction time to see if there is a pattern or trend. Better still, you can plot an x-y graph of say, age against reaction time. If the sample includes adults, you should find a very significant pattern.

	A	B	C	D	E	F	G	H
1	**Reaction time**							
2	**NAME**	**Height**	**Specs**	**Age**	**Best 100m**	**Reaction time**	**Reaction time**	**Sex**
3	Sertac							
4	Sonia							

You can go on to study how quickly we can stop a car. See the Forces section under *Friction: Braking distance*.

Key stage 3 Handling information using a Spreadsheet

Using IT in ... environmental and pollution studies

Do a survey of the class.

There are endless questions you might ask about human variation. Who has the biggest hands? Who is the tallest? Do tall people have bigger feet? Can taller people jump higher? Can they swim further? Does the length of your legs, your stride or the size of your feet help you to sprint faster? Can shorter people balance for longer on a tightrope or broomstick? Does chest size affect lung volume? Why can some people throw a ball further than others? Is it because of their height? Or their arm length? Or the flexibility of their shoulder joint?

A **spreadsheet** or **database** program can help you record the results of such surveys and answer the questions too. It's important to start by establishing the questions you wish to answer. It's also important to agree on what data you are going to collect and how it will be recorded - for example, you might agree to measure length in cm.

> **Pupil Worksheet**
>
> **See the Database topic**

When you've entered the data you can analyse it. You might first sort the individuals into order - for example, by the pupils' height. Next, you might plot bar graphs of pupils heights; plot pie charts of shoe sizes and plot x-y graphs to look for patterns between two variables. There is enormous scope here for creativity - you might collect data on gender, age, hand span, stride length, reaction time, tongue-rolling, ear free or attached, leg length, athletics results, chest size, lung volume, lung pressure or pulse rate before and after exercise. There are many interesting patterns to discover, for example, you might find that girls have smaller feet for their height than boys. Or for something more focused, try to find out 'Which aspect of the size of your hand affects how much sand you can hold?'. You weigh handfuls of sand collected by each member of the class and measure their hand spans, wrist sizes, finger length or whatever to find which measurements correlate best.

	A	B	C	D	E	F	G
1	**Class survey**						
2	NAME	Colour of Hair	Colour of Eyes	Height cm	Weight cm	Shoe Size	Sex
3	Sertac	Brown	Black	129	29	3	Boy
4	Sonia	Blue	Brown	130	33	3	Girl
5	Geoffrey						
6	Cara						

Key stage 3-4 Handling information using a Database program or Spreadsheet. See School Science Review Mar 91

Explore and run a nature reserve.

You can get software to simulate the running of a nature reserve and gain some experience you might never get otherwise. For example, using **Ecodisc** - an interactive adventure on CD-ROM (Mac - AVP/ Macademic) you can take a trip round a nature reserve and make observations. You can see the reserve in winter and summer, make crucial management decisions and see the consequences of your decisions almost immediately.

Key stage 3-4 Modelling using Interactive multimedia / simulation program

Worry about the entire planet.

SimEarth (PC - AVP) allows you to take charge of the earth from its birth. You guide the development of life from microbe to civilisation, control the atmosphere and volcanoes, track continental drift and biological diversity. The program is quite detailed and yet this 'whole earth model' can make difficult ideas accessible. There's more 'geography' here than science but the program is impressive.

Biosphere (CD-ROM PC - Education Interactive) is a detailed encyclopaedia of the environment. It covers green issues and ecology. It needs teaching materials and adds only a little to a book on the subject. More so for the **Big Green Disc** (CD-ROM PC - Education Interactive).

Key stage 4-A level Modelling using a simulation program

Why do plants grow in different places?

Investigate places where plants grow by using sensors to measure the features of different environments. You'll find sensors to measure the **pH** of soil, the **temperature**, the **wind speed** and the **light level**. You can record the readings in a **spreadsheet** and use it to prepare graphs.

Key stage 3-4 Measuring using sensors

Using IT

Section 3

Using IT in ... environmental and pollution studies

Do some plants grow better than others at a particular site?

Do a survey of plant life across a section of land. Organise the results as a table and enter them into a spreadsheet. The program can help by doing calculations of areas and averages. It can plot the results on pie or bar graphs.

	A	B	C
1	**Where do plants prefer to grow?**		
2	**Plant**	**Site A**	**Site B**
3	Grass		
4	Daisy		
5	Yarrow		
6	Buttercup		
7	Plantain		
8	Clover		

For something more structured, i.e. designed for surveys of this type, use **Ecosoft** (BBC / Arc / Nimbus - AVP). More up to date is Fieldworks (PC - Interpretive Solutions)

Key stage 3 Handling information using a Spreadsheet

How much of soil is water? How much is organic matter?

Pupil Worksheet

See the Spreadsheet topic

Measure the content of a number of soil samples - by weighing and heating dishes of soil in the usual way. Pupils can be helped to do their calculations by using a **spreadsheet**.

	A	B	C
1	**Water content**	**Result 1**	**Result 2**
2	Mass of dish g	250	
3	Mass of dish and wet soil g	355	
4	Mass of wet soil sample g	105	
5	Mass of dish and dry soil g	301	
6	Mass of dry soil sample g	51	
7	Mass of water g	54	
8	Percentage water	51%	

Key stage 3-4 Handling information using a Spreadsheet

Ecology II

Study the rain forest and ...

A field trip to the Rain forest is a pictorial simulation of the floor of the rain forest (Mac - TAG). There are pictures of animals and plants to click on and find out more about. It includes data tables and food chains for pupils to compile. It's suitable for pupils from 7 to 15. There's also a companion program, **A field trip to the sea,** which is a pictorial simulation of the kelp forest. It includes a food chain activity which 'challenges pupils to identify predator/prey relationship in the forest'. (Mac - TAG). See also: Habitats.

Key stage 3 Modelling using a Simulation program

Population ecology

A very manageable model of a population is **Creatures** (age 14-18, PC disc from Future Skill Software). This shows the interdependance of foxes, rabbits and plants. For an advanced level simulation, see **Population ecology** (Mac /PC - TAG). This is a dynamic ecology simulation. You can design a plant or an animal population, food webs and environments with varying habitats and physical barriers. You can then track changes in the population levels, biomass and distributions.

A level Modelling using a Simulation program

How do populations of wolves and deer interact?

Pupil Worksheet

See the Spreadsheet topic

The interrelationships of two populations is a good example of where the computing and graphing features of a **spreadsheet** can be very useful. See the worked example on wolves and deer.

	A	B	C	D	E	F
1	**Wolves and deer**					
2		Deer alive at the start of the year	Deer died of sickness or old age	Deer killed by wolves	Deer born this year	Deer alive at the end of year
3	1970	1000	100	100	205	1005
4	1971	1005	95	110	215	1015
5	1972	1015	110	105	200	1000
6	1973	1000	110	115	205	990

Key stage 3-4 Modelling using a Spreadsheet
Idea from Salter's Science

Using IT in ... environmental and pollution studies

Environment: waste

Do a waste survey.

Do a survey of the waste we produce. Find out what sorts of things we throw away and how much of our rubbish is recyclable. Use a **spreadsheet** to record the data. The spreadsheet can draw a pie graph to compare the relative amounts of the things we throw away. Using the data it can calculate how much of our rubbish is recyclable and again plot this on a pie chart.

How has our dustbin changed over the years?

You'll often come across data handling exercises in science schemes. In one example, pupils had to draw graphs to compare the average dustbins of today with ten and forty years ago. Using a **spreadsheet** is entirely appropriate here - the program can make light work of graphs and allow us to spend more time analysing the data.

	A	B	C	D
1	**Has history changed our rubbish bins?**			
2	Item	40 years ago	10 years ago	Today
3	Card / paper	8%	25%	30%
4	Dust	75%	4%	0%
5	Glass	5%	10%	10%
6	Metal	3%	7%	10%
7	Organic waste	3%	38%	30%
8	Other	5%	10%	12%
9	Plastics	0%	5%	8%

Pupils can use the results of their own waste survey, and you can 'create' typical dustbins for them to study. They can go on to suggest how and why our waste has changed.

Key stage 3-4 Handling information using a Spreadsheet
Idea from Salter's Science

Write a letter to the newspaper to encourage care for the environment.

You can use a **word processor** to write a letter. The report can make a series of practical suggestions to encourage people to do something about the environment. The word processor provides an excellent medium for pupils to work collaboratively.

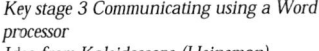

Key stage 3 Communicating using a Word processor
Idea from Kaleidoscope (Heineman)

Habitats for animals

What sort of invertebrates live in the leaf litter of a wood?

Questions such as this are good starting points for a survey. You can record how many of each type of creature live in leaf litter or under a tree canopy. You then enter the data into a **spreadsheet**. The program can prepare the graphs you want to draw. **Ecosoft** (BBC / Arc / Nimbus - AVP), is a program specifically designed for environmental surveys of this kind and should be helpful here. More advanced and detailed is **Fieldworks** (PC - Interpretive Solutions).

Key stage 3-4 Handling information using a Spreadsheet
Idea from Information Technology in Science (MEU Cymru)

How are habitats different?

Use a **graphics program** to draw a plan of a habitat. Mark features about the area such as dampness, temperature, light level and soil pH. Show where the plants live and annotate the plan saying how the environment might affect life within the habitat. For example, say how street lighting or human activity disturbs it.

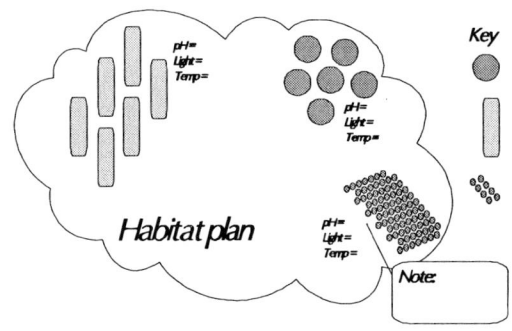

Key stage 3 Modelling using a Graphics program
Idea from Bath Science (Nelson)

Habitats to explore.

PictureBase Habitats illustrates and documents the many habitats in the natural world. This is a large photographic and text database covering over 20 different habitats - from wetland to roadside. (Disc or CD-ROM PC / Arc - AVP). Microsoft's **Ocean** (CD-ROM PC - TAG) makes a excellent library title. Other possible library titles include **Barrier Reef** (CD-ROM PC - AVP), **Oceans Below** and **Amazon Trail** (CD-ROM PC / Mac - AVP). **Eyewitness Encyclopaedia of Nature** (CD-ROM PC / Mac - mail) does habitats and food webs well, though it's hardly an encyclopaedia. Both **Sonoran Desert** and **Worlds of the Reef** (PC CD-ROM for age 12+ , Ransom) are quite special, detailed studies of habitats.

Using IT

Section

3

Using IT in ... environmental and pollution studies

How do rock pools change during the day?

You can use **temperature** and **conductivity sensors** to measure the salinity of water in rock pools. Attach the sensors to a meter or data logger to take readings. The data logger can help by taking readings over a long period of time.

Key stage 3-4 Measuring using sensors

Pollution I

How does the acidity of rain affect the corrosion of marble?

You can measure the rate of carbon dioxide evolved from acid and marble chips by using a gas syringe. If you connect a **position sensor** to the syringe, the sensor arm will be moved by the plunger and you can monitor the evolution of carbon dioxide. You should be able to obtain some excellent graphs to show say, the effect of acid concentration on gas evolution. You might investigate whether the pH of 'rain' has anything to do with gas evolution.

There is a '**Key**' database file called **Acid Drops** (All machines - AVP). Dated in looks, this contains the results of a national survey into acid rain with data from rain samples taken across the country. This throws up just a few interesting patterns - in particular you can plot the pH results on a map and see the spread of the results.

Key stage 4 Measuring using sensors / Handling information using a Database program

Make a poster about ozone depletion or acid rain.

A **word processor** or **graphics program** can help pupils produce realistic posters. They can work together to write the text, they can even use scanned-in images to illustrate it.

Key stage 3 Communicating using WP / DTP programs

Water pollution

Use an **oxygen sensor** to measure the oxygen level of some water samples. The change in oxygen level can be used to work out the biological oxygen demand or BOD.

Fishkill (PC/Nimbus - AVP) is for a slightly older group, say aged over 14. With this you can take on the role of a pollution officer working for a river authority. There might be an 'incident' where dead fish are found in a part of a river. You then have to 'take' water samples, analyse them and investigate the source and nature of the pollution. This is an above average problem solving exercise, it is old but the idea of playing detectives is gripping enough.

A level Measuring using sensors / Modelling using a simulation program.

Pollution II:

How does our use of fuels affect atmospheric carbon dioxide?

Textbooks provide tables of data about our use of fuels and our production of carbon dioxide over time. It's a simple matter to enter that data into a **spreadsheet** table and plot it on a graph. The pupils can plot an x-y graph to find if there is a pattern between the use of fuel and production of carbon dioxide. They can easily extrapolate their graph to see when our production of carbon dioxide will say, double.

	A	B	C	D
1	How much CO2 do I produce?			
2	Quarter	Gas Therms	Electric kWh	Car litres
3	Spring	0	0	0
4	Summer	0	0	0
5	Autumn	0	0	0
6	Winter	0	0	0
7	Totals			
8	CO2 per unit	5	1	2.5
9	Total CO2			

Key stage 3 Modelling using a Spreadsheet

Using IT in ... materials

How do petrol and diesel oil compare for pollutants?

A **spreadsheet** is well suited to activities involving numerical data and graphs. It cuts short the less important activities such as drawing graphs and leaves more time for interpreting the data. The starting point is a table of data you might find in a science book - in this example, about the pollution caused by petrol and diesel oil. Enter the data into a spreadsheet and draw pie and bar graphs to compare the fuels.

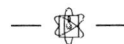

You might ask for example, what would happen if we used exclusively one fuel or the other. It is an easy matter to calculate this using the spreadsheet and to plot a new set of graphs to find out.

	A	B	C
1	**How do petrol and diesel engines compare?**		
2	**Pollutant**	**Petrol engine**	**Diesel engine**
3	Carbon monoxide	280	8
4	Nitogen oxides	25	12
5	Hydrocarbons	16	25
6	Solids	2	16
7	Sulphur oxides	1	5
8	**Total**	324	66
9	**All in g/litre fuel**		
10			
11			
12	8% 12%		
13			
14	24% 18%		
15			
16			
17	Diesel engine		
18	38%		
19			

Key stage 3-4 Handling information using a Spreadsheet
Idea from Folens Copymasters

Key stage 3-4 Handling information using a Spreadsheet
Idea from Folens Copymasters

Acids

Draw a pH indicator colour chart.

Use a **drawing program** to draw 14 boxes and number them from 1 to 14. Add the words acid, very acid, alkaline, very alkaline and neutral. Colour in the boxes with the colours corresponding to universal indicator.

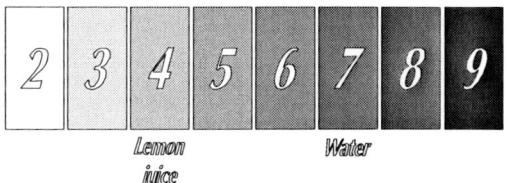

Add an example of each type of solution for example, 'Lemon juice', to the chart.

Key stage 3 Communicating using a Graphics program

Atomic structure

Draw the electron shell diagrams of the first 20 elements.

Use the circle drawing feature of a **graphics program** to create electron shell diagrams.

See the periodic table CD-ROM software titles, listed under Metals and Non-metals, for some good models of atomic structure. One especially relevant title is **Atom Viewer** (PC/Mac CD-ROM from New Media) - an on-screen exercise where you have to fill the correct number of electons into each shell of the first twenty elements.

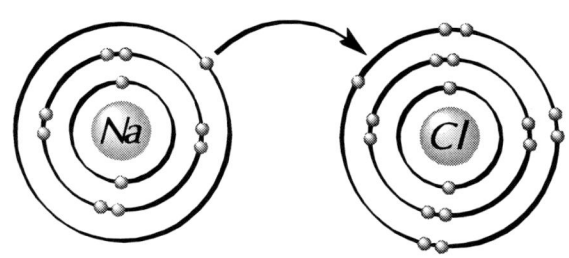

Key stage 4 Communicating using a Graphics program.

Using IT

Section

3

Using IT in ... materials

Separating mixtures

How can we separate salt from salt and sand?

An idea for a cut out exercise is to get the pupils to sort the steps for a method into order. Using a **word processor** you can considerably shorten the time usually spent on this. Create a word processor file with say, the steps for separating salt from sand. Ask the pupils to use the word processor to rearrange the steps on the screen into the correct order.

Pupil Worksheet
See the Word Processor topic

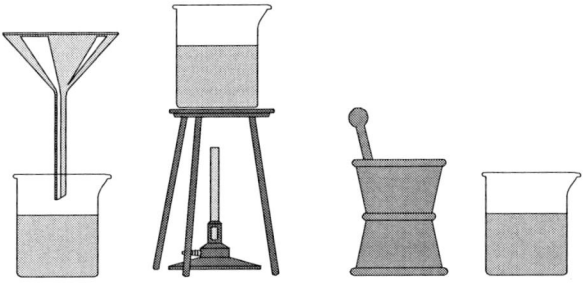

You can go a bit further and add a list of the reasons for each step and ask for these to be put into place too. The finished piece can then be printed.
Key stage 3 Communicating using a Word processor

Change of state: heating curves

What happens to the temperature when a substance changes state?

Measure the temperature of a cooling substance such as wax or stearic acid. Use a **temperature sensor** to measure and plot the temperatures on a graph. It's also well worth measuring the temperature as you heat a beaker of ice - the graph is quite impressive for such a simple experiment. Stirring it continuously helps.

An interesting variation of the experiment is to place a test tube of the cooling substance in a lagged container of water. You then use sensors to measure its temperature, as well as that of the water. The cooling substance will show the usual stepped graph, but the temperature of the jacket of water rises, and continues to rise even when the substance is changing state.
Key stage 3-4 Measuring using sensors

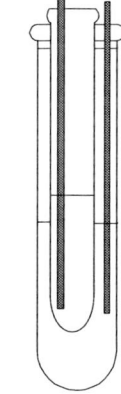

Elements, mixtures and compounds

See **Elements, compounds and Mixtures** (PC/Mac - CD-ROM, New Media) is a likeable multimedia tutorial on this topic. Content is also relevant to the curriculum.
Key Stage 3

Chemical formulae

Write word and symbol equations...

Word processors have unique formatting abilities which make writing equations much easier. For example, using a table to write the equations is a tidy improvement on writing them by hand. The word processor will also allow you to use subscripts (e.g. 'H_2O') for chemical formulae. For a novel exercise based on this idea, use the word processor to rearrange a number of mixed-up chemical equations.
Key stage 3-4 Communicating using a Word processor

Learn the chemical symbols and chemical formulae.

Use a drill and practice program, but appreciate that they offer nothing to teaching science processes. Nevertheless formulae do have to be learnt.
Key stage 3-4 Communicating using a drill and practice program

What is the formula of magnesium oxide?

Weigh magnesium ribbon in a crucible before and after burning. Enter the class' results into a **spreadsheet** and use it to calculate the mass of MgO. Plot an x-y graph to show the combination ratio of Mg to O.

	1	2	3	4	5	6	7	8
1	What is the formula of magnesium oxide?							
2	Group	Mass of crucible	Mass of cru' + Mg	Mass of cru' + MgO	Mass of Mg	Mass of MgO	Moles of Mg	Moles of O
3	1	20	21					
4	2							
5	3							
6	4							
7	Average							

You can get a clear view of the class' results - seeing which results are spurious.
Key stage 4 Handling information using a Spreadsheet

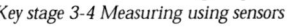

Using IT in ... materials

Density

Compare the densities of materials.

Compare the densities of metals, woods, rock, oil, water and sand. To give this some focus you might ask which timber is the most dense? Or how do the densities of bird bones and human bones compare? Either way, first measure the mass and volumes (or length, breadth and height) of a number of materials, then enter the results into a **spreadsheet**. The spreadsheet can calculate the densities for you. Use the spreadsheet to sort the materials into order, and then draw a bar graph to compare them.

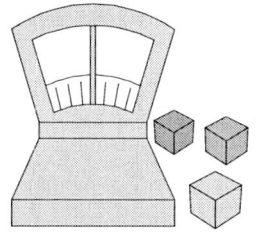

	A	B	C	D	E	F	G
1	Comparing densities of materials						
2	Number	Material	Length	Breadth	Volume cm3	Mass g	Density g / m3
3	1						
4	2						
5	3						
6	4						
7					Exercise		
8					Volume cm3	Mass g	Density g / m3
9					10	80	?
10					?	200	20
11					20	?	1.5
12					10	8	?
13					?	12	1.2

You can also set up a worksheet exercise on density which the pupils complete at the computer. Type in a spreadsheet table with mass, volume and density data. Remove one value from each row and ask the pupils to use formulae to calculate the missing values.

Key stage 3-4 Handling information using a Spreadsheet

Expansion

How much do things expand?

A **position sensor** can be used to illustrate the expansion of materials as it can measure their change in shape. You might show the expansion of a balloon of air, a bar of metal or a part-filled gas syringe. Warm the materials with the lever arm place of a position sensor resting on a balloon, metal bar or syringe plunger. Similarly, allow them to cool.

Key stage 3-4 Measuring using sensors

Materials: classification

Can you sort the materials into sets?

You can use a **drawing program** to create a novel cut-out exercise. You'll need a collection of clip-art or scanned-in pictures of about 15 different objects - made from various materials.

Arts & Letters Editor

Using the mouse the pupils can sort the materials into sets - hard and soft; flexible and stiff; clear or opaque or some other property. Each sorting exercise can be labelled and printed in turn.

Key stage 3 Handling information using a Graphics program
Idea from Science Scene (Hodder)

Build a key to identify different materials.

Collect a set of materials - elements, laboratory chemicals or household materials for example. Sort them into types - paying attention to say, their colour, texture, solubility or natural/man-made origins. Then use a **branching database** program to create a key to identify them.

See the Branching Database topic

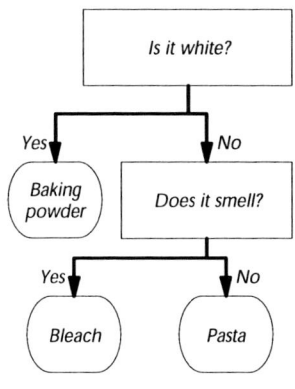

The program helps you to structure a series of questions about the materials so building up a useful key. When you've finished, test the key with other members of the class and make changes where the questions are perhaps ambiguous.

Key stage 3-4 Handling information using a Branching Database program

Using IT

Section

3

Using IT in ... materials

Materials: uses

Which material would you use for...?

Choosing the right materials will help ensure things last as long as they should. What properties would you want in the material for the roof of a garden shed? What about a ski jacket, a car body or a child's bath toy? What metal would you use to make the wiring for a fire sprinkler?

To answer these questions, you will need a **database** of materials and their properties.

Comparing materials							
Material	Colour	Hard or soft	Brittle or flexible	Shiny or dull	Conducts electricity	Heavy or light	Absorbent or water proof
Glass							
Polythene							
Lead							
Copper							
Pottery							

You search the database to find the material you need. You could make your own database - and involve pupils in an interesting research activity. Otherwise you could use a ready-made database. For example, **Materials** is a database you use with the **Key** data handling programs (PC / Arc - SCSST). Here you'll find information on wood, plastics, ceramics, textiles and metals - all indexed by name and properties. Another **Key** database covering this area is **Materials, components and techniques** (All - Anglia). There's some science value in this one, but it's mostly for technology. There is more science in **Materials** (CD-ROM PC / Arc - Granada) - a bank of information on 150 diverse materials. A few nice touches - like testing a material for elasticity but on its own this demands little student involvement and is largely decorative. **PictureBase: Materials** (CD-ROM PC / Arc - AVP) is a bank of pictures and text on the variety and behaviour of materials.

Key stage 3 Handling information using a Database program

How has our use of metals changed?

Which metal shows the biggest decrease in use over the years? The answer is easier to find when you enter the data into a **spreadsheet**. You can use the program to quickly prepare graphs that allow you analyse it.

	A	B	C	D
1	How has our use of lead changed?			
2	Use of lead	1960	1980	2000
3	Alloys with other metals	12	5	3
4	Batteries	28	51	64
5	Electrical cable	19	8	2
6	Paint pigment	10	14	13
7	Petrol	8	6	4
8	Roof sheeting	14	8	7
9	Various	9	8	7
10	TOTAL	100	100	100

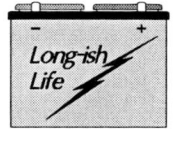
Long-ish Life

Key stage 4 Handling information. From Blackwell Modular Science

Metals and non-metals

What patterns are there in the properties of the elements?

Pupil Worksheet
See the Database topic

Do metals melt at high temperatures? How do their properties (melting point, conductivity, crystal structure) compare with non-metals? Which elements have a melting point over 400°C? Which elements have a boiling point over 800°C. Which elements are good conductors? Which are good insulators? Which can be bent? Which can be hammered? You can search the data in a ready-made **database** of the elements. They contain more data than you can normally collect together. It is also very easy to sort the data and look for patterns. The

See the Branching Database topic

Key data file on the **Periodic Table** (Arc / PC - AVP) is used with the **Key** family of data handling programs. With this you can sort the elements by group or melting point. You can plot an x-y graph of melting points against atomic number. You can study a single group or the whole table. **The Chemistry Set 2000** (CD-ROM PC / Arc - New Media) is the best yet software for looking for periodic patterns, melting points and discovery dates of elements. There is masses of information and a section where you can watch many chemical reactions. It is, however, what you can make of it - you'll certainly need to make some classroom activities. New Media have an Internet linked version of this called **Multimedia Chemistry School**. This should be seen. **Elements** (CD-ROM Arc / PC - YI) is more decorative than analytical, it shows off many aspects of the periodic table with photographs, diagrams, graphs and film but alarmingly no serious pattern searching is possible here. Beware **Interactive Periodic Table** (PC - Attica) - this is pretty poor: it is weak on searching for periodic patterns and there is little 'chemistry' here.

Key stage 4 Handling information using a Database program

Record No.
Data search

Using IT in ... materials

Materials: absorbing water

Which material will absorb the most liquid?

Compare materials such as cloths, sponges and nylon to see which will absorb the most liquid. You'll need to compare equal or at least measured quantities of material (either by weight or area). You might choose to weigh the materials before and after a timed period of wetting - but either way there are plenty of variables to control. Put the results into a **spreadsheet** to calculate the weight of water absorbed and the amount of water absorbed per quantity of material. You'll be able to do the calculations with ease and indeed make the exercise accessible to younger scientists.

	1	2	3	4	5	6
1	**Which is best to soak up water?**					
2	Material	Dry weight	Wet weight	Amount of water	Amount of material	Water absorbed per amt.
3	J cloth					
4	Nylon					
5	Newspaper					
6	Tissue					

Or do an intriguing experiment to find out which material is the best for waterproofing. Wrap a sample of material (say, nylon, silk or cotton) around a **humidity sensor**. Place this in a poly-thene bag with a measured amount of water and a **temperature probe**. Monitor the temperature and the humidity over time. Then repeat with the other fabrics in turn.

Key stage 3-4 Handling information using a Spreadsheet / Measuring using sensors. Idea from the Softlab software documentation.

Other materials

What liquids are used to run a car?

Cars need all sorts of liquids to run them. They serve functions with plenty of scientific interest. Ask the class to make a table to list the liquids, their functions and where they are 'stored'. Normally, drawing tables is a bit of a pain but making one on a **word processor** is easy and tidy. Word processor tables are flexible - they let you squeeze in new ideas as they arise. This IT idea is simple and flexible and given good access to technology it can help improve the quality of pupils' work.

Key stage 3 Communicating using a Word processor Idea from Science Scene (Hodder)

Build a key to identify different plastics.

Collect of set of plastics. Sort them out focusing on how they behave - how they melt, whether are heat-stable, flexible or brittle. Next use a **branching database** program to create a key to identify the plastics. The program helps you to structure questions and build up a key.

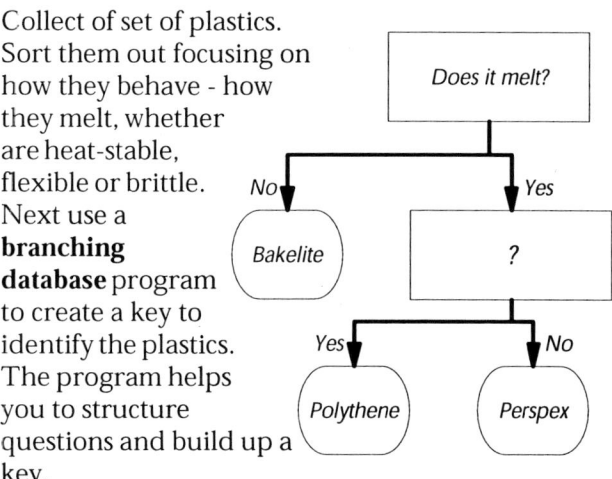

Key stage 3-4 Handling information using a Branching Database program

Should we ban plastic disposable goods?

"A member of Parliament is heard on the radio saying that plastic disposable goods should be banned. Write a letter to the MP explaining the advantages of plastics". This exercise, involving some extended writing to summarise a topic of work, is a good place to use a **word processor**. The pupils work in pairs and use the word processor to write a letter to the MP. The gist of their letter might be that while they understand the problems caused by plastics, there are good reasons why they shouldn't be dismissed lightly.

Key stage 3 Communicating using a Word processor Idea from Science Scene (Hodder)

Using IT in ... chemistry / physical changes

Dissolving

How can we get gelatine to dissolve faster?

Investigate the effect of temperature on the time taken to dissolve a piece of gelatine. Investigate the effect of surface area on dissolving gelatine. Use a **spreadsheet** program to record the results. Plot an x-y graph of time taken to dissolve against surface area.

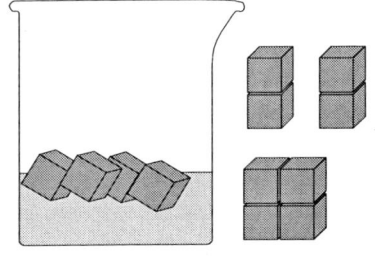

	A	B	C	D	E	F
1	How does surface area affect dissolving?					
2	Test	Length	Width	Depth	Volume	Time to dissolve
3	A	4	2	1	8	
4	B	3	2	1		
5	C	2	2	1		
6	D	1	2	1		

Key stage 3-4 Modelling using a Spreadsheet
Idea from the Essex spreadsheet posters (Essex)

Evaporation

Why do dogs pant on a hot day?

There's plenty of scope for investigating evaporation using **temperature sensors** which measure and display temperature changes. The visual feedback from the graph can be a great aid to understanding as well as more successful experiments. Find out for example, why dogs pant on a hot day or why athletes pour drink over their heads after a race. Or even why wet gloves feel so much colder? Does aftershave really make the skin cold or does it just feel cold? Is there any connection between how well aftershave works and its alcohol content? And so on.

Key stage 3-4 Measuring using sensors

Mountaineers take care to keep dry, why might this be?

Getting clothes wet can not only make us feel cold, but it is also dangerous. Does the wind affect how we survive the cold? Can waterproof materials help us? To investigate this cover two cans of hot water with anorak material - one of course, will be wet. Measure the rate at which they cool using **temperature sensors** - the sensors seem to exaggerate the smallest temperature changes. Monitor the temperature as two cans of warm water, one insulated and one not insulated, cool down. You can use an electric fan to simulate a cold wind. Repeat the experiment but cover one of the 'anoraks' with plastic - this prevents the evaporation of water from the wet anorak material.

Key stage 3 Measuring using sensors

Gases

Why is heating used batteries dangerous?

> **Pupil Worksheet**
>
> See the Spreadsheet topic

To illustrate how pressure changes with temperature, carefully heat a closed container of air using a water-bath. Take readings of the pressure at various temperatures. Use a **spreadsheet** to record the results of this experiment. Get the program to draw an x-y graph of temperature against pressure.

For modelling at an advanced level, the simulation program **Gas Laws** (Mac/PC - Explorer, TAG) can be used here.

Key stage 4 Modelling using a Spreadsheet

	A	B	C	D	E
1	Boyle's Law - the experiment				
2					
3	Pressure x100,000N/m2	Volume cm3	PxV	1/Volume	
4	3.13	16.0	50.08	0.0625	
5	2.97				
6	2.78				
7	2.56				
8	2.42				
9	2.21				
10	1.97				
11	1.83				
12	1.60				
13	1.40				
14	1.20				
15	0.95				
16					
17					
18					
19					
20					
21					

Graph of Volume against Pressure for a fixed Mass of gas

Using IT in ... chemistry / physical changes

How does pressure change with temperature and volume?

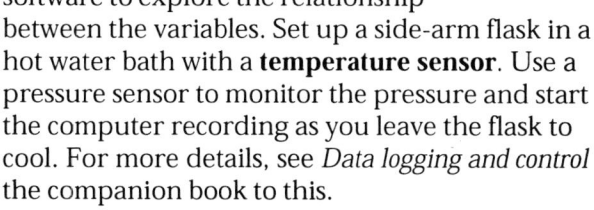

You can measure temperature, volume and pressure using computer sensors. It's good in that you can see the changes taking place on a graph as they happen. You can also use the software to explore the relationship between the variables. Set up a side-arm flask in a hot water bath with a **temperature sensor**. Use a pressure sensor to monitor the pressure and start the computer recording as you leave the flask to cool. For more details, see *Data logging and control* the companion book to this.

Key stage 4 Measuring using sensors

What is the best temperature to keep drink fizzy?

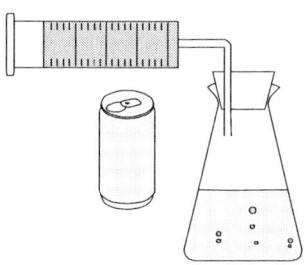

Get a flask of cold drink and attach it to a gas syringe. Warm the flask slowly and record the volume of gas evolved using a **pressure sensor** attached to the flask. You could also use a gas syringe and rest the lever arm of a **position sensor** on it. A **temperature sensor** can record the temperature change. Ordinarily this is a very fiddly experiment but using sensors allows us to monitor two changing quantities and to plot them on a graph simultaneously.

Key stage 3-4 Measuring using sensors

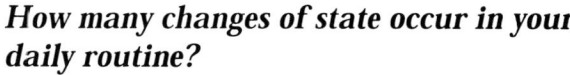

Kinetic theory

How are solids liquids and gases different?

Moving Molecules (BBC - CUP now deleted) and **States of Matter** (Mac/PC CD-ROM - New Media) offer very useful simulations of our model of kinetic theory. You can use them to explore what happens to particles when you increase the temperature. Recommended.

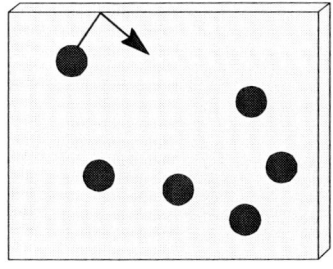

Key stage 3-4 Modelling using a Simulation program

Is everything a solid, a liquid or a gas - or are there exceptions?

You can use the **word processor** to write, make tables or prepare posters. Efforts on paper tend to be messy but their real weakness is that they can't be improved in the way that words on the screen can. In this example the pupils make a table of things which don't fit into our solid, liquid, gas classification. They build up their table with explanations of why these things are exceptions - if indeed they are.

As an alternative, the pupils might use the program to make a poster explaining how things change state.

Key stage 3 Communicating using WP / DTP
Idea from Kaleidoscope (Heineman)

How many changes of state occur in your daily routine?

Use a **word processor** to write a story listing all the changes of state that occur in a daily routine. For example, you get up on a cold morning, see condensation on the window, take a bath, make coffee, sprinkle salt on the icy door step... By using the word processor the pupils gain the ability to develop their 'story' - adding new ideas as they come to mind. When they've finished they can read through the passage and highlight (for example, underline, change to bold face or italics) the words that relate to a change of state.

Alternatively, the teacher could write the story and save it on disc. The pupils could then work on the story, finding the changes of state and sorting them into lists - for example, 'Gas to Liquid' might be one list. Unlike work with pen and paper there's no cutting or copying out - and more time can be spent on-task.

Key stage 3 Communicating using a Word processor
Idea from Science Scene (Hodder)

Using IT in ... chemistry / physical changes

Does the temperature change steadily when you heat ice?

Use a **temperature sensor** to measure the temperature as a beaker of ice is heated. Stir it continuously. As you measure you'll see a graph on the screen and pupils can try to predict where the graph trace will turn next. Will it follow a straight line? Why does the graph stop rising when the water is boiling? Print the graph and annotate it with the reasons for each turn of the graph.

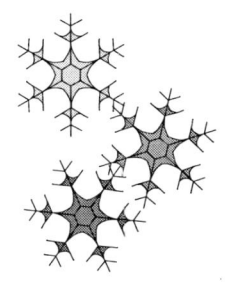

Key stage 3 Measuring using sensors

Molecular models

Building molecular models

Molecular modelling programs - programs allow you to construct and see spatially-correct molecules on-screen. In **Nemesis Sampler** (Mac/PC - Oxford Molecular) you build a molecule from a vast library of atoms and groups. You can rotate the molecule in space, get it to change to its lowest energy conformation and do several other advanced manipulations. This 'sampler', costs a fraction of the full price but lacks the ability to print. **Desktop molecular modeller** (PC / Arc emulator - Philip Harris) is less restricted and better than equal. **Arachne Moleculer Modeller** (Arc floppy, from TAG) allows you to build molecular models - though that is not much more than real physical models. Search the web for **Rasmol** which does lots and for free - and look out for **Molecules 3D** (www.molecules.com). Some graphics programs feature chemistry clip-art. You can use these pre-drawn chemical models for your worksheets. Some programs also have dazzling drawing features which help you to draw 3D models. For example see **Micrografx Designer** (PC Windows) which can produce this kind of specimen:

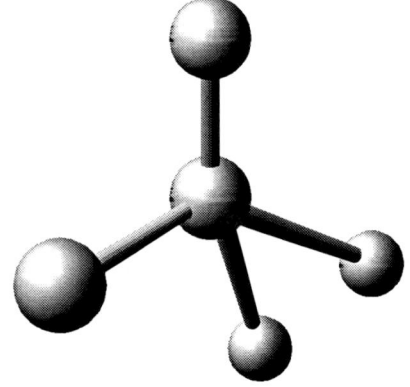

Using IT

Section

3

A level. Modelling using a modelling program

Radioactive decay and penetration

Will nuclear waste stay with us forever?

You can enter results from a radioactive decay experiment into a **spreadsheet**. The spreadsheet will allow you to extrapolate, or work out when you could expect to reach a certain count rate. It could tell you how much activity there would be after 10 half lives or how long a source would take to reach background level. The spreadsheet can show, from one line to the next, how the count rate halves with each half-life period. It can total the half-lives to show for example, what difference changing the half-life value from say, a minute to a million years would make.

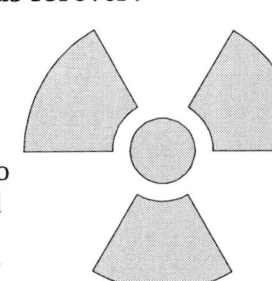

	A	B	C	D	E F G H I J
1	Radioactive decay - theoretical				
2	Time units	Count	Half-life	Time units	
3	0	10000	0.75	hours	
4	0.75	5000			
5	1.5	2500			
6	3	1250			
7	6	625			
8	12	313			
9	24	156			
10	48	78			
11	96	39			
12	192	20			
13	384	10			
14	768	5			
15	1536	2			

Note that you can also use **a spreadsheet** to find the dose of radioactivity you would get from exposure to various radioactive sources. This too is a good example of modelling using IT.

Key stage 4 Modelling using a Spreadsheet

What is the half life of a radioactive material?

Use a **radioactivity sensor** to monitor the rapid decay of radioactive sources such as Protactinium. The decay curve, which you obtain 'as it happens' is unusually good and a great aid to understanding. You can also use the 'as it happens' display to help demonstrate the penetration of materials by alpha, beta and gamma rays. The **Radiation Game** (UKAEA) provides an introduction to this area of knowledge.**Science Series: Elements** (CD-ROM PC/Arc - Granada) has some good simulations of radioactive decay. See also Newbyte's titles at www.newbyte.com

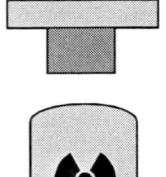

Key stage 4 Measuring using sensors / Simulation program

Using IT in ... topics on chemical change

Burning

Is oxygen used up in burning?

To illustrate the idea that oxygen is used up in burning, you can burn a candle in a jar and use an **oxygen sensor** to measure the oxygen level. You'll see a novel graph of the oxygen level falling and reaching a plateau. For a bit more interest, re-admit air into the container when the candle extinguishes and continue measuring for a short while. Incidentally, if you have a **humidity sensor** you can simultaneously measure humidity - providing some evidence of water production in burning.

Key stage 3 Measuring using sensors

How long can a candle burn for in a closed container?

It's interesting to see how the size of a container affects how long a candle burns for. Time how long a candle burns for under differently sized beakers. Enter the results into a **spreadsheet** and use the program to plot an x-y graph to find the relationship between volume and time. A straight line graph should help us to extrapolate. You might ask "suppose you didn't have a 2L beaker to test, how long would the candle burn for?"

	A	B	C	D	E	F
1	How long will the candle stay alight?					
2	Volume cm3	1st go	2nd go	3rd go	Average	
3	100	2.29	1.92	1.69	1.97	
4	150					
5	250					
6	400	From MEU Cymru				
7	500					

Key stage 3-4 Modelling using a Spreadsheet
Idea from Information Technology in Science (MEU Cymru)

Chemical equilibrium

Equilibrium.

A simulation program, **Chemical Equilibrium** (Mac/PC - Gateways - TAG) provides a particularly comprehensive look at this. It allows a good degree of control and visual feedback. You can choose the reactants, alter the activation energy or the rates of the forward and reverse reactions, simulate the drop by drop addition of chemicals, in, say, an acid-base titration, label atoms in a molecule and trace its path in a reaction. Though the topics are different, **Electrochemistry** and **Chemical Kinetics** (Mac/PC - Explorer, TAG) are in the same series, at same level and may also be of interest. **Gas Equilibrium** (PC - Newbyte) allows you to experiment with factors affecting equilibrium.New Media have a very good title covering the **Haber Process** (PC/ Mac - New Media)

A level Modelling using a Simulation program

Concentration

Which bleach is the best value for money?

Get a number of fresh branded bleaches and record their cost and sizes. You can test their bleaching power by counting the number of drops of ink (or blue food dye) a small sample can decolourise. The results of the experiment are handled by a **spreadsheet** program. Using it you can easily calculate the cost per cm^3 and the bleaching power of each brand.

	A	B	C	D	E	F
1	Which bleach is the value for money choice?					
2	Bleach	Bottle size	Cost per bottle	Cost /100cm3	Drops of ink	Cost / ink bleached
3	Domestos					
4	Tesco					

Key stage 3-4 Modelling using a Spreadsheet
Idea from School Science Review Sept. 88

Using IT

Section

3

Using IT in ... topics on chemical change

Make a calibration curve.

A **spreadsheet** is an ideal tool to create a calibration curve for chemical analysis.

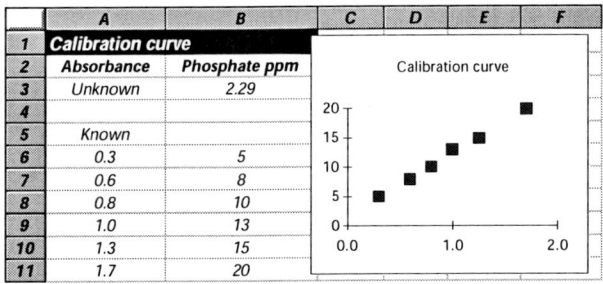

	A	B	C	D	E	F
1	Calibration curve					
2	Absorbance	Phosphate ppm				
3	Unknown	2.29				
4						
5	Known					
6	0.3	5				
7	0.6	8				
8	0.8	10				
9	1.0	13				
10	1.3	15				
11	1.7	20				

A level. Modelling using a Spreadsheet. See School Science Review Dec. 92

Endo- and exothermic reactions

How does the temperature change when ammonium sulphate is added to water.

Use a **temperature sensor** to measure the temperature change when chemicals such as sodium carbonate or ammonium sulphate are added to water. If you are demonstrating such temperature changes you'll find the large computer display particularly useful, if not essential.

How is the temperature change related to surface area?

When sodium carbonate is added to water heat is produced. With a couple of **temperature sensors** you can show the effect of surface area on the rate of heat production. For example, you can compare adding large crystal and powder sodium carbonate to water. A graph on the screen will show, to good effect, how powdered chemical produces the heat fastest.

What is the best mixture for a mountaineer's heat-pack?

Use a **temperature sensor** in an unusual investigation to find the best mixture for a heat pack. The idea is to produce a mixture which will give out heat for the longest time. You have to experiment with the quantities of lime, icing sugar and water to find the optimal mixture. Similarly, you can use an endothermic reaction to design or study a 'sports injury' pack.

Key stage 3-4 Measuring using sensors

Using IT

Section

3

Food processing

What is the best mixture for making jam set?

This interesting idea for an investigation involves using different amounts of sugar in a jam mixture. You dissolve sugar in water and boil to make a gel. You then add pectin and allow the mixture to cool. You repeat with different amounts of sugar and test the viscosity of each jam by timing its flow through a filter. Finally you enter the data into a **spreadsheet** and use this to plot a graph. You should find a relationship between the viscosity of the jam and the sugar concentration.

	A	B	C	D
1	Which recipe makes a thicker jam?			
2	Amount of sugar	Time to filter - secs		
3	5	3.5		
4	10			
5	20			
6	30			
7	40			
8	50			
9	60			
10	70			
11				
12				
13				

Key stage 4 Modelling using a Spreadsheet. Idea from Mike Hammond's Handling data using Spreadsheets and Databases (Sheffield University)

How does cooking affect the vitamin C level in peas?

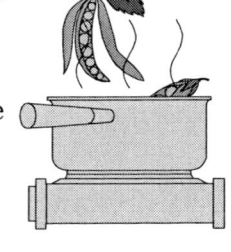

Boil some peas for 30 minutes. Every few minutes take a sample and analyse its vitamin C level. Enter the results into a **spreadsheet** and plot an x-y graph of vitamin C level against time. What advice would you give to the cook?

	A	B	C	D	E	F
1	How can we preserve vitamin C in cooked food?					
2	Cooking time	Remaining vitamin C/100g				
3	in Minutes	Beaker X	Beaker Y	Beaker Z		
4	2	60	60	60		
5	4	55	48	33		
6	6	50				
7	8	45				
8	10	41				
9	12	37				
10	14	34				
11	16	31				
12	18	28				
13	20	25				
14						
15						
16						
17						
18						

Key stage 4 Communicating using a Spreadsheet. Idea from Salter's Science

Using IT in ... topics on chemical change

Lattice Energies

What patterns are there in the lattice energies of the alkali metal halides?

You can model the lattice energies of the alkali metal halides using a **spreadsheet**. You enter the details from a data book and use the spreadsheet to help with the calculations.

	A	B	C	D	E	F	G	H	I
1	Modelling lattice energies								
2	Substance	Charge a	Charge c	Ionic radius a	Ionic radius c	Inter-atomic distance	Force/ N	Ion energy	Tear apart energy
3	NaCl								
4	NaBr		From the 'IT in science pack' from MEU Cymru						
5	NaI								
6	LiCl								

A level Modelling using a Spreadsheet
Idea from Information Technology in Science (MEU Cymru)

Electrochemistry

What happens in electrolysis?

A very polished cartoon tutorial can be found in **Electrochemistry** (age 14-15, CDROM for Mac/PC from New Media). This also looks at the electrolysis of bauxite for aluminium.
Key stage 4.

Metal extraction

Where do metals come from?

Do a data search to find out where the metals come from. You can use a ready-made **database** on disc or CD-ROM. These contain far more data than you can normally collect together.
You can sort the data and find say: which metals come from the sea; which metals come from rocks or which metals must be extracted using electricity. See for example, the **Key** file on **Rocks & Minerals** (All machines - AVP) or the disc on the **Elements** (CD-ROM PC/Arc - Granada).

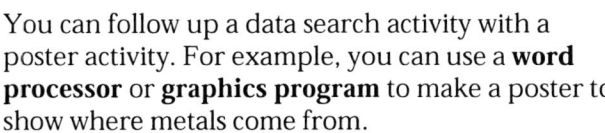

You can follow up a data search activity with a poster activity. For example, you can use a **word processor** or **graphics program** to make a poster to show where metals come from.
Key stage 3-4 Handling information using a Database program. Communicating using word processing / DTP programs

Rates of reactions

What affects the rate of a reaction?

A simulation program, **Rates of Reactions** (PC- New Media) allows you to investigate the effect of different factors on the rate of a reaction. The program animates the reaction collisions.
Key stage 4 Modelling using a Simulation program

How does the acid concentration affect its rate of reaction with marble?

You can measure the rate of carbon dioxide evolved from acid and marble chips by using a gas syringe. If you connect a **position sensor** to the syringe, the sensor arm will be moved by the plunger and you can monitor the evolution of carbon dioxide. You should be able to obtain some excellent graphs to show say, the effect of acid concentration on the rate of gas evolution.
Key stage 4 Measuring using sensors

How does temperature affect the reaction between acid and sodium thiosulphate?

A **light sensor** or **colorimeter sensor** can measure the progress of the reaction between acid and sodium thiosulphate.
Like a colorimeter, it monitors the light transmitted through the mixture over time. The results and graphs can be quite convincing.
Key stage 3-4 Measuring using sensors

What affects the reaction between bromine and methanoic acid?

Computer sensors can monitor the above reaction and present a graph of its progress in 'real time'. With bromine and methanoic acid you use a **light sensor** or **colorimeter sensor** to measure the light transmitted through the solution over time.

How do different catalysts affect the decomposition of hydrogen peroxide?

Computer sensors can monitor the decomposition of hydrogen peroxide. You can use a gas syringe and connect a **position sensor** to it. As oxygen is given off, the gas syringe plunger moves the sensor arm. (Use a **pressure sensor** instead if you have one). To increase the range of catalysts you might use celery, liver as well as manganese dioxide.
Key stage 4 Measuring using sensors

Using IT in ... topics on chemical change

Do a survey to show which cars suffer the most rust.

Computers are good at recording and helping you analyse data from surveys. In this example, you record the age of each car and devise a way of recording how much rust each car has. Enter the data into a **spreadsheet** or **database** program and use it to sort and graph the data.

	A	B	C	D	E	F	G
1	Do some cars rust more than others?						
2	Car	Reg letter	Year	Rust on visible areas	Rust on front panel	Rust on doors	Total score
3	Orion	D	1988	2	3	0	5
4	Audi	D	1988	0	0	0	0

Key stage 3-4 Handling information using a Spreadsheet or Database

Titrations

How do strong and weak acids compare?

Use a **spreadsheet** to record the results of a titration. As usual, you record the volumes and pH in titrations between alkali and strong / weak acids. You use the spreadsheet to plot graphs of pH against volume and so compare the equivalence points of strong and weak acids.

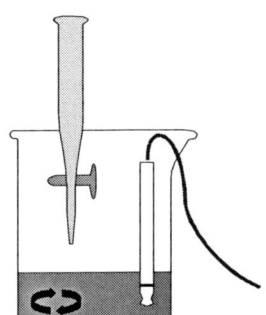

You could also use a **pH sensor**. The sensor can monitor the pH automatically - you only need to type in the volume of acid you have added. As with the exercise above, you'll end up with a graph of pH against volume. It is even more interesting to monitor the temperature at the same time - you'll be able to see the heat of neutralisation reach a peak at the equivalence point. Do remember to use slightly more concentrated solutions to get a reasonable response. See also www.newbyte.com for a neat titration simulation.

	A	B	C	D	E
1	Strong acid - strong base titration				
2	Acid volume cm3	Strong base pH	Weak base pH		
3	0				
4	2				
5	4				
6	6				
7	8				
8	10				
9	12				
10	14				
11	16				
12	18				
13	20				
14	22				
15	24				

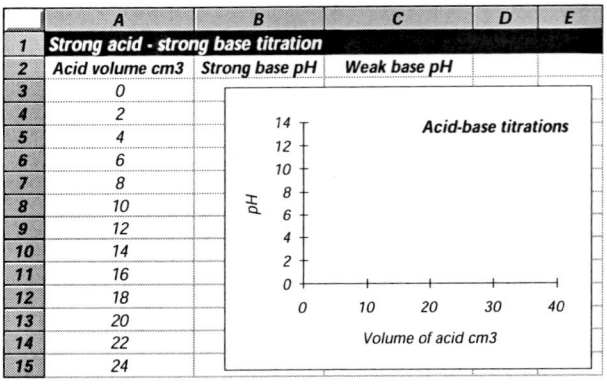

Key stage 4/ A level. Handling data using a Spreadsheet

How does the conductivity of a solution change during a precipitation reaction?

Use a **conductivity sensor** to take readings and plot a graph for you during a conductimetric titration.
A level Measuring using sensors

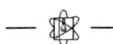

Using IT in ... atmosphere and weather

Air

Draw a graph to show the proportions of the gases in the air.

Enter the amount of each gas in the air into a **spreadsheet** - just as you would fill in a table. Use the program to plot a bar graph and a pie graph. Which graph is the most appropriate? If you wish, the graphs, good and bad, can be included in a word processor report.

	A	B	C	D	E
1	**What makes air?**				
2	**Gas**	**%**			
3	Helium	0.005			
4	Carbon Dioxide	0.03			
5	Others	0.065			
6	Argon	0.9			
7	Oxygen	21			
8	Nitrogen	78			
9	TOTAL	100			

Key stage 3 Handling information using a Spreadsheet.
Idea from Kaleidoscope (Heineman)

Are some rooms more humid than others?

You can use a sensor to measure the **humidity** of the air in different parts of the school. You will need to attach the sensor to a meter or data logger to take the readings. You could also measure the temperature at the same time - to find a pattern between humidity and temperature levels.

Key stage 3-4 Measuring using sensors

How can you make the washing dry more quickly?

Wet a piece of fabric and hang it on an **electronic balance** linked to the computer. The balance can measure the rate at which the washing dries. Use a fan to help dry the fabric - use it at different settings to see how the 'wind' speed affects drying. You will obtain a picture of the rate at which the material dries, whereas normally you would not get the same feel of how fast things dry. Repeat the experiment, drying other materials.
It would be interesting to see if you could relate the **humidity level** (measurable with a sensor) to how fast things dry.

Key stage 4 Measuring using sensors
Idea from Blackwell Modular Science

Icy weather

Why do they put salt on the roads?

Use **temperature sensors** to measure the temperature of melting ice, salt and ice and sand and ice at the same time. You should gain a useful set of graphs to compare side-by-side.
You can also measure the extent to which salt depresses the freezing point. You will need a fridge and two temperature sensors. Place both sensors in an ice cube tray - one dipping in plain water, the other dipping in salty water. Leave to freeze while you monitor their temperatures. You should get some results within half-an-hour.

Key stage 3-4 Measuring using sensors

Rocks

Which rocks give us useful materials?

Do a data search on a ready-made database of rocks. You might use the **Key** data file, **Rocks and Minerals** (All machines - AVP). Use the **Key** program to search and sort the rock database. You can find out which rocks can be used in their native state; which rocks require heating to obtain useful materials and which require electrolysis. A nice feature of this data file is that it can show the distribution of rocks on a world map. Using this feature you can see if any parts of the world are particularly rich in useful rocks. See also: Earth Sciences.

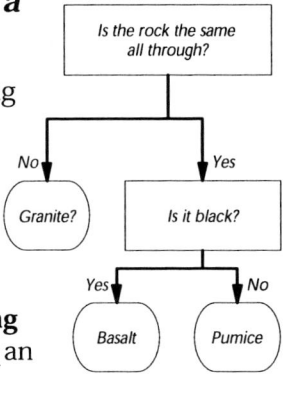

Key stage 3-4 Handling information using a Database program

Create a key to identify a set of rocks.

Sort out a set of rocks paying special attention to their colour, their hardness, the size of their crystals, whether they are conglomerate and if they contain any useful ingredients. Use a **branching database** program to create an identification key. The program helps you to structure the key and can be the centre of an engaging observation exercise.

Key stage 3-4 Handling information using a Database program
Idea from Information Technology in Science (MEU Cymru)

Is the rock the same all through?
No → **Granite?**
Yes → **Is it black?**
Yes → **Basalt**
No → **Pumice**

Using IT in ... atmosphere and weather

Water

How much water do you use?

Keep a record of how much water your household uses in a day. Enter the data into a **spreadsheet** table and use it to total your water use. Calculate how much water you use in a month or in a year. If you like, convert the amounts to bath-fulls. Experiment with the figures to see how you could economise on water. Use the spreadsheet to draw a bar graph or a pie chart comparing our different uses of water.

	1	2	3	4
1	**How much water do we use?**			
2	Use	Amount of water	Number of times	Water used
3	Shower	10	3	
4	Basin wash			
5	Loo flush			
6	Wahing up			
7	Laundry			
8				
9	TOTAL			

See also **Water** (PC / Arc CD-ROM - Granada) for its broad look at water usage and its implications. Historical and geographical data is included here and it overshadows some of the science.
Key stage 3 Modelling using a Spreadsheet

Write an account of a water particle as it travels round the water cycle.

Reach for the **word processor** whenever the task involves an extended piece of writing. Arrange the group so that the pupils work together.
Key stage 3 Communicating using word processing / DTP programs
Idea from Kaleidoscope (Heineman)

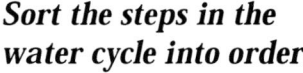

Sort the steps in the water cycle into order.

Before the lesson, type in the steps in the water cycle and save the work on disc. Then ask the pupils to use the **word processor** to help them sort the list into a correct order. They can change the sorted list into prose and print out their work.
Key stage 3-4 Communicating using a Word processor

Using IT

Section

3

Label a diagram of the water cycle.

Create a diagram of the water cycle using a **graphics program** or **scanner**. The pupils can use the program to label the diagram. Their labels can describe each step in the water cycle and they can annotate the diagram with words such as rising, falling, evaporating, cool, heat, condense.

Key stage 3 Communicating using a Graphics program
Idea from Oxford Science Programme (OUP)

Weather

Monitor the weather.

Pupil Worksheet
See the Database topic

Use sensors or a dedicated **weather station** to monitor the weather. Keep cuttings from the newspaper to cross-check the measurements. You can usually transfer the weather data to a **database** or **spreadsheet** where you can turn the data into graphs. There are numerous patterns to be found and using 'real' data makes the effort more meaningful.

There are now devices which allow you to pick up weather maps from the **satellites** in space. These, **remote sensing** devices let you see a snapshot of the world's weather as it happens. They also provide views of the world through different cameras or filters. For example, there is an infra-red map showing hot and cold areas and you might track the temperature changes in the Sahara desert throughout the day. They are certainly an excellent context for developing scientific knowledge - about satellite orbits, the weather, signal transmission and modern technology. It would be an excellent idea to acquire such a system, but how it should fit into the curriculum needs to be taken to the geographers - as should **Understanding the Weather** (PC for age 16-19 from Granada) and **European Weather** (for PC from SCET).
Key stage 3-4 Measuring using sensors Blackwell Modular Science

Using IT in ... atmosphere and weather

How is wind speed affected by pressure?

Use a **pressure sensor** to measure the atmospheric pressure and a **rotation sensor** to measure the wind speed over a few days. You should find a pattern between the two measurements. You might want to connect them to an automatic data logger to avoid tying up the computer. Compare your findings with those on a newspaper weather report. You might also look for a pattern in the isobars when the wind is strong.

Key stage 4 Measuring using sensors
Idea from Science Scene (Hodder)

Make a weather report for a newspaper.

Use a **graphics program** to draw a set of weather symbols. Use a scanner to 'grab' a map you could use for a weather report. The pupils can then arrange the symbols to create a weather forecast for a newspaper. They can annotate the map with an explanation of their forecast.

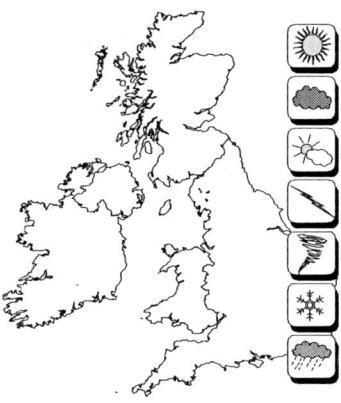

Key stage 3-4 Communicating using a Graphics program
Idea from Kaleidoscope (Heineman)

Look for patterns in the weather

Pupil Worksheet
See the Database topic

Create a weather **database** - using data from the newspaper, from a dedicated weather station or the Meteorological office. Use the data to find out if the following ideas are true. It gets cold before it rains. The wind blows faster on cold days. Winter months are the rainiest. August is the warmest month. It rained more last year. Of course, you will need to plan ahead for this activity but the subject is forever throwing up interesting patterns and challenging our ideas.

Key stage 3 Handling information using a Database program
Idea from Handling data with a Spreadsheet (Sheffield University)

More weather

Weather Mapper (Mac / PC / Arc - TAG) is an above-average and easy picture book about the weather. You can also use it to record your own weather data, display it on a graph or map and make your own weather report. It also carries a small database of world weather trends.

	A	B	C	D	E	F	G	H	I
1	**What is weather?**								
2	Date	Sunshine	Wind direction	Wind speed	Pressure mb	Relative Humidity %	Rain fall mm	Max temp	Min temp
3	6.694	2	NW	7	1015		5	21	18

Key stage 3 Modelling using a Simulation program

Weather across the world

How do rainfall and temperature vary across the world?

Numerous questions arise from comparing weather data across the world. For example, you might ask what is the difference between the maximum and minimum temperatures in each country? Which country has the biggest difference? How might this affect the life in those countries? In which country would it be the most uncomfortable to live? In which place would it be the most difficult for animals to survive?

You can enter the weather data into a **spreadsheet**. The spreadsheet can, for example, calculate the differences between the maximum and minimum temperatures in each country. It can draw bar charts to compare rainfall and temperature in different parts of the world. It can also draw a combined bar graph of highest and lowest temperatures.

	A	B	C	D	E
1	**Weather and climate**				
2	Country	Rainfall	Max temp	Min temp	Temp difference
3	UK	2	21	18	
4	N.America				
5	Sahara				

If getting weather data is a problem see the **Key database** file (All machines - AVP) on **Weather and Climate**. This is a file you can use with one of the **Key** database programs. It contains weather data from around the world at different times of the year. A nice bonus is that you can say, search for all the hot places and see them plotted on a world map. See also **Weather Mapper** (see above) which is more structured and more exciting too.

I have also heard of colleagues using electronic mail to find out about weather across the world. They exchange messages with schools using their computer via a modem. As you might not expect, this can prove to be enjoyable and absorbing.

Key stage 3-4 Handling information using a Spreadsheet / Database

Using IT in ... electricity and magnetism

Wind: effects on structures

How does the wind affect different structures?

Build a set of structures and then attempt to damage them using the wind or more practically, a fan. There are many variables to control here and so it makes a worthy investigation. To make the exercise more quantitative, you can use a **rotation sensor** to measure the wind speeds achieved

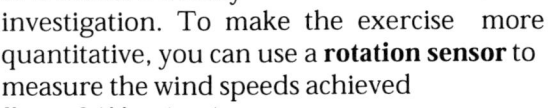

Key stage 3-4 Measuring using sensors

Advanced electricity topics

Exploring electrodynamics

See the **Electrodynamics** simulation (Mac /PC Windows - TAG). This substantial program allows you to investigate the effects of a uniform electric current or magnetic field on a charged body. You can explore the drift path and velocity of a particle in a mixed electric/magnetic field. You can also perform activities that simulate the Milikan oil drop experiment, a cathode ray tube, a cyclotron and a bubble chamber. This is above average and should be seen.

A level Modelling using a Simulation program

Exploring electrostatics

Electrostatics is a model with plenty to explore (Mac /PC - TAG). It allows you to position charged particles or specific shapes and then display the charge distribution and equipotential lines. You can show forces between the charges and the charges can be allowed to move under the forces generated. You can also investigate Coulombs law, a Faraday cage and a field near a sharp object. Should be seen.

A level Modelling using a Simulation program

Measure the induced current as a magnet falls through a coil.

Use a sensor to measure the change in **potential difference** as a magnet drops through a flat circular coil. You will need to get the computer to record as fast as possible. This experiment provides something which is otherwise very difficult to see.

Key stage 4 - A level. Measuring using sensors

Explore the magnetic field along the axis of a coil.

Use a **magnetic field sensor** to take a series of readings at different positions along a coil.
For a computer based 'model', see **Electromagnetism** (PC - Visual Products). This is an exceptional, advanced program where you move an iron loop within a magnetic field and examine its effects in detail.

A level Measuring using sensors

Capacitor discharge

What affects the discharge of a capacitor?

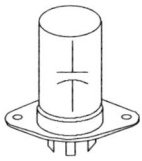

A capacitor is a device which stores electrical energy as electrical charge. If the capacitor is discharged, the charge leaks away and as it does, the voltage across it drops. You can use a spreadsheet to model the loss of charge from a capacitor. In other words - the spreadsheet can represent the rate of discharge both numerically and graphically. You enter the decay formula into the spreadsheet and you can then change the time, the value of the capacitor, the value of the resistor and see how these affect the discharge. You can also of course, capture real data from a discharging capacitor using **current** and **potential difference sensors** or a **Capacitor Module** (Deltronics)

	A	B	C	D	E
1	**Capacitor Discharge**				
2	Enter or change the following details:				
3		Capacitor value	500	microfarads	
4		Discharging Resistor	100000	ohms	
5		Charging potential	10	volts	
6		Time steps	2	seconds	
7		Number of steps	30		
8					
9	Time	Charge	Potential	Current	Change in
10	s	microC	volts	mA	charge microC
11	0	5000	10	0.1	200
12	2	4800	9.6	0.096	192
13	4	4608	9.216	0.09216	184.32
14	6	4423.68	8.84736	0.0884736	176.9472
15	8	4246.7328	8.493466	0.084934656	169.869312
16	10	4076.863488	8.153727	0.08153727	163.0745395

Key stage 4 / A level. Modelling using a Spreadsheet / Measuring using sensors

Using IT in ... electricity and magnetism

Chemical energy: batteries

Compare batteries to see how long they last.

Use a **potential difference** or **voltmeter sensor** to take measurements in a circuit with a battery and lamp. Let the computer or data logger take the readings for you as you leave this running. You might try this with Ni-Cads, alkalines, zincs or a lead-acid accumulator. The results take a while but the results justify the effort. You can use the results to help you select the best type of battery for a toy car, a cassette recorder, an electric toothbrush or an automobile. You can also explore how well batteries recover after a period of use.

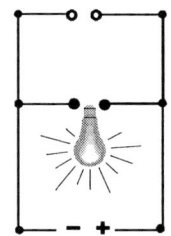

Key stage 3-4 Measuring using sensors. See School Science Review Sept. 90

Circuit diagrams

Drawing circuit diagrams.

Draw a circuit diagram for a table lamp using a

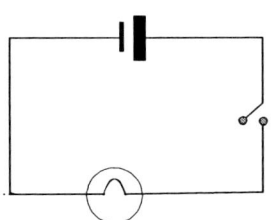

graphics program; draw the circuit for the switch system on a staircase. Draw a circuit for a motor racing set where you can control the speeds of the cars. Draw two (i.e. series and parallel) circuits for Christmas tree lights and explain how they are different. What would happen when a bulb in the set blows?

By using a graphics program you can assemble such circuits from symbols which you have previously stored on disc. To draw a circuit, you load a symbol from the disc and join them up with lines. A reasonably IT capable group should be quite productive in this sort of exercise.

Key stage 3-4 Modelling using Graphics programs

Communication

How are communication signals transmitted around the world?

If you have access to an **Internet service**, this can be a starting point for explaining the technology for communication. Some schools have set up IT projects with other schools and used a **fax machine** to exchange information - the BP/ASE Science Across Europe project does this.

Key stage 3-A level Modelling using a Simulation program / Communicating using IT

Control

Build a burglar alarm.

A simple and sensible approach to create a burglar alarm system would be to solder together some components or fit together electronics modules. However, should you wish to illustrate our ideas about **computer control**, your computer sensors can be put to use to create an equivalent, if expensive, computer controlled system. For a sensor you might use a pressure switch mat (which the burglar treads on) or an infra-red sensor (which responds to body heat). For an alarm use a flashing lamp, a strobe light, a buzzer or a siren. You connect these up to a **buffer box** and write a simple program using **control software**.

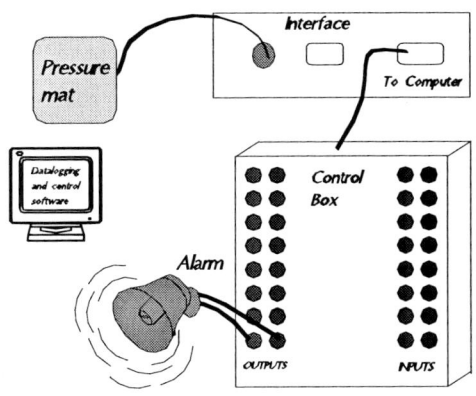

Key stage 3-4 Measuring using Control technology

Using IT in ... electricity and magnetism

Electrical energy: domestic use

How much electrical energy do you use at home?

Do a survey of the power of electrical appliances in the home. Estimate how long you use each appliance for each day. Enter the data into a **spreadsheet** program and use the program to estimate the electricity bill for a day, a week or a month. You can use the completed spreadsheet to estimate the bill at other times of the year and to find ways to economise on electrical energy.

	A	B	C	D	E	F
1	How much electricity do we use?			One unit of energy costs p		6
2	Appliance	Power W	Power kW	Time used hours	Energy units kWh	Cost of electrical energy p
3						
4	Computer	150	0.15	6	0.9	5.4
5	Cooker					
6	Electric clock					
7	Electric light					
8	TOTALS					

Key stage 4 Modelling using a Spreadsheet

Electricity models

There are programs which model our use of electrical energy. They can show us how to economise on the use of electrical energy. **At Home with Wattville** (BBC / Arc / PC - UE) has merit - it

looks at meters, energy saving and how we use electricity. For example, it shows what happens to the energy bill as you turn various appliances on.

Key stage 3 Modelling using a Simulation program

Design the lighting for a building to ensure that energy is not wasted.

The switches and the positions of the lamps in a building affect the way we use electricity. For example, you should be able to switch off lights near a window. Similarly attention to insulation and other energy saving features will help towards a successful design. Use a **graphics program** to position the lights and switches on a plan of the building. Use the program to annotate the plan showing all the energy saving features of your design.

Key stage 3 Modelling using Graphics programs

How can we keep electricity safe?

The drill / tutorial program, **Fuses** (PC / Arc - UE)

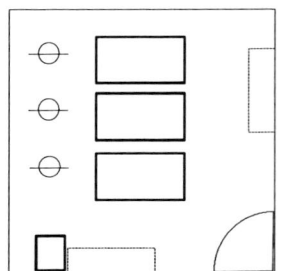

allows you to investigate fuses. It covers the function of the earth wire, domestic circuits and the idea of a ring main. An easy, colourful look at a not too practical subject.

Key stage 3 Modelling using a Simulation program

Electrical effects

Examine the effects of an electric current flowing through a wire.

When electric current flows through a wire you can find a heating effect (which we can use to keep warm) and a magnetic effect (which we can use to drive electric motors). You can use sensors to measure either of these. To study the heat produced by a current, set up a low voltage heating coil in a small beaker of water. Use a **current** or **ammeter sensor** to measure the current and a **temperature sensor** to measure the temperature of the water. The computer shows the temperature change as the water is heated. You can then study the effect of different currents.

You can similarly use a **magnetic field sensor** to

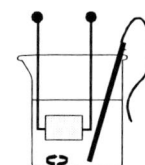

measure the strength of the magnetic field inside a coil of wire and study the effect as you change the current in the wire. You can then get the computer to plot the field strength against the current.

Key stage 4 / A level. Measuring using sensors Blackwell Modular Science

Using IT

Section

3

SOFTWARE and CDROM REVIEWS can be found in "SOFTWARE FOR TEACHING SCIENCE" © IT in Science

Using IT in ... electricity and magnetism

Electricity and electronics

Exploring electricity and electronics.

Crocodile Clips (Mac/PC - from Croc Clips) is an excellent circuit diagram drawing and modelling tool for serious work on electronics. It's hard to choose between it and **Edison** (age 8-18, PC - Quickroute) as both are very good. The latter can also be used throughout the secondary school. **FutureLAB Circuits** (age 11-18 CD-ROM for PC from Nicholl Education) is a photorealistic simulation to use as an electronic blackboard. **Electronics Workbench** (PC / Mac - TAG) needs skill to exploit its versatility while **Quickroute** (PC - Quickroute) is an advanced drawing tool for printed circuit design. **Oak PCB** (Arc - TAG) is less glamorous but no less valuable. Before you make the real thing you can test your circuits on screen using **Oak Logic** (Arc - TAG). For advanced work there's **Logicworks** (Mac/Arc - AVP) and **AC/DC Circuits** (Mac/PC - TAG). For electronics theory there is the tutorial-style **Basic Electronics** (CD-ROM PC - BTL). **Electricity and Magnetism: Science series** (Acorn/PC, for age 11-15, Granada) is a fair look at electricity in the home, fuel sources and experiments. **Electricity and Magnetism** (CD-ROM PC - AVP) is an 'electronic blackboard' covering including circuits and logic gates. **Animated Circuits for Education** (PC CD-ROM - AVP) I wasn't animated by these last two. **The Electronics Series** (PC CD-ROM for age 15+ , from Matrix) has interactive labs and information on circuits, components and symbols. **The way things work 2** (CD-ROM PC / Mac - mail) shows how everything from the battery to the television works. It includes facts which are brilliantly cross referenced - a good title for project work in the home and library. **Inventors and Inventions** (PC / Arc - Granada) - it is a mine of information, but just another picture book.

Key stage 4 to A level. Modelling using a Simulation program

Electronics

How a thermistor responds to temperature

Set up a circuit with a power supply and a thermistor in a beaker of hot water. Place a **temperature sensor** in the beaker and **connect voltage** and **current sensors** into the circuit. Allow the water to cool as the computer records. Work out the resistance of the thermistor at various temperatures. Plot a graph of resistance against temperature and note the lack of straightness of this graph. Advanced students can plot the logarithm of resistance against 1/temperature - which should now produce a straight line.

Key stage 4 - A level. Measuring using sensors

What is the effect of light on a light dependant resistor?

Take a series of readings in a circuit containing an LDR. Use sensors to measure the **light level**, the **current** and the **potential difference**. Use the sensor software to plot a graph of resistance against light level. Alternatively, use conventional meters to take the readings and enter the results into a **spreadsheet**. Use the program to plot an x-y graph of resistance against light level.

	A	B	C	D
1	How does an LDR respond to light?			
2	Light level	Current	Voltage	Resistance
3	0			
4	10			
5	20			
6	30			

Key stage 4 Handling data using a Spreadsheet

Cost your electronics projects.

In a project, such as a design & technology project to make an electronic device from several parts, there are numerous different resources to quantify and cost. For example, you might be making a quantity of flashing badges for a charity fair. If you are, it's very easy to put the figures into a **spreadsheet** - as a table of parts, numbers and unit costs. You can then work out the cost of the materials. You can use the spreadsheet to experiment with the quantities - working out how many items you need to sell to make a profit.

	A	B	C	D	E	F
1	How much will my project cost?					
2	Item	Cost per pack	No. per pack	Cost each	No. per product	Cost per product
3						
4	LED	£5.00	5	£1.00	2	£2.00
5	Battery					
6	Lead					
7	Solder					

Key stage 3-4 Modelling using a Spreadsheet

Magnets

Make a poster about magnets for an exhibition at the science museum.

Use a **word processor** and to prepare the text for a poster about magnets. Say what magnets are, what they are made of and how they are used in everyday devices. If you have the facility, use a **graphics program** to assemble the poster. You might even use computer **'clip-art'** to illustrate it.

Key stage 3 Communicating using word processing / graphics programs
Idea from Kaleidoscope (Heineman)

Using IT

Section

3

Using IT in ... electricity and magnetism

Ohm's law

Ohm's law calculations.

In electricity topics we often ask pupils to 'work-out' resistance using current and voltage values. Instead you can enter the figures into a **spreadsheet** and use this to present the exercise. You make up a spreadsheet with columns for current, voltage and resistance and then save the file. The pupils use the spreadsheet to work out the answers.

Ω

Omega

	A	B	C	D	E
8	**Resistance calculator**				
9	Item		Current	Voltage	Resistance
10	Lamp		0.5	12	24
11	Toaster		3	240	80
12	Cake mixer		2.2	240	109

Key stage 4 Handling information using a Spreadsheet

How does the current through a device change when the voltage is changed?

Set up a circuit with a power supply, rheostat and lamp. Instead of conventional meters, connect sensors instead - you can get sensors which measure **potential difference and current**. Move the rheostat slider and the computer will take all the voltage and current readings in just a few seconds. Repeat the exercise, replacing the lamp with a resistor or a diode.

Ω You will be impressed by the ease and simplicity of the **Current-Voltage Module** (Deltronics). It allows you to explore a wealth of electrical measurements using a computer. You can use it to compare different resistors, to compare a resistor with a bulb and to investigate the power dissipated in each case. You can also compare the characteristics of silicon and light emitting diodes or the output of a transistor. In fact there's a whole booklet of investigations **Electrical Measurements** (Roy Barton - UEA/Leicester) to take you through, step by step.

You could use conventional meters to take readings in the same sort of circuit - i.e. one with a power supply, rheostat and lamp. This time you enter the readings into a **spreadsheet** (see above). Use the spreadsheet to both plot an x-y graph of **v i** and to calculate the resistance.

Using IT

Section

3

Key stage 3-4 Modelling using a Spreadsheet / Measuring using sensors

Resistance graphs and calculations

How does the length of a wire affect the resistance? Measure the current passing through lengths of Constantin wire. Enter the results into a **spreadsheet** and calculate the resistance of the wire. Plot an x-y graph of length and resistance. What is the difference between series and parallel circuits? Measure the current passing through two resistors, first arranged in series and then in parallel. Enter the results into a spreadsheet and use the program to help explain the difference between the two sorts of circuit.

	A	B	C	D	E
1	**How does the length of a wire affect resistance?**				
2	Length of wire	Material	Current	Voltage	Resistance
3	10	Constan			
4	20				
5	30		3		
6	40				
7	50				

Key stage 4 Modelling using a Spreadsheet

Resistance of soil: applications

Produce a contour map to find the buried treasure.

In the search for old ruins, archaeologists can discover buried objects by taking point-to-point soil resistance readings. They then plot the readings on a contour map - the resulting peaks on the map can indicate buried items. The plotting alone takes an inordinate amount of time. Instead, you can enter the data into a **spreadsheet**. The spreadsheet will be able to draw a 3D surface graph which will show the resistance contours very quickly.

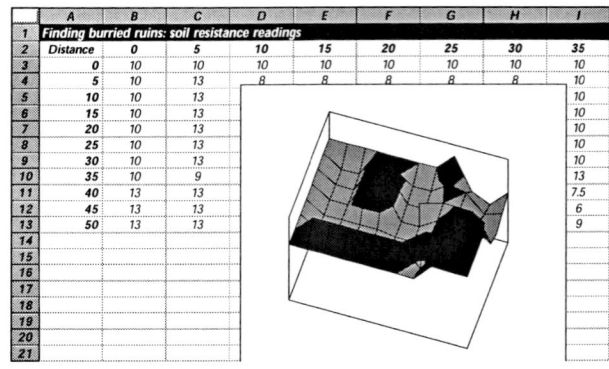

	A	B	C	D	E	F	G	H	I
1	**Finding buried ruins: soil resistance readings**								
2	Distance	0	5	10	15	20	25	30	35
3	0	10	10	10	10	10	10	10	10
4	5	10	13	8	8	8	8	8	10
5	10	10	13						10
6	15	10	13						10
7	20	10	13						10
8	25	10	13						10
9	30	10	13						10
10	35	10	9						13
11	40	13	13						7.5
12	45	13	13						6
13	50	13	13						9
14									
15									
16									
17									
18									
19									
20									
21									

Key stage 4 Modelling using a Spreadsheet
Idea from The Physical World (Nelson)

Using IT in ... Energy

Sensors - what they do

How can sensors help us?

Place different **sensors** around the room as an exhibition or circus. Make up question cards for each sensor. The cards might ask: what makes this sensor respond? What part of this is the sensitive part? You might also provide a list of everyday uses of sensors and ask the pupils to match them to the different types of sensor.

Key stage 3-4 Measuring using sensors

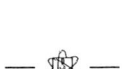

— ⚛ —

Alternative energy

Make and test a solar cooker.

You can make a solar cooker from the parabolic reflector of a glow-bar electric fire. Or you might use an umbrella lined with shiny foil. Use a **temperature sensor** to find the hot spot of the solar cooker and measure its temperature over a period of time. Could you use this for cooking? You may need to use a data logger if you wish to take readings for any length of time or if you are far from a power point.

Key stage 3 Measuring using sensors

Make a device which swivels towards the sun to maximise the light it receives.

You can make a computer controlled, sun-seeking collector. You will need two **light sensors** and a motorised platform to turn the collector to face the sun. It need not be particularly elaborate - but you will need a control buffer box. You then use **control software** to write a 'program' which compares the readings from the sensors. If one

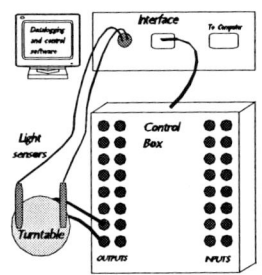

sensor receives more light than the other, you can work out which direction the sun is coming from. The program can then rotate the solar collector until the two readings are equal.

Key stage 3-4 Measuring using control technology

Could the sun supply our electricity needs?

Set up a circuit with a solar cell. Use sensors to measure the **current**, the **potential difference** and the **light** level. See how the current produced varies throughout the day.

Key stage 3-4 Measuring using sensors

How do water wheel designs compare?

There are several different designs of water wheel. For want of better names, there is the pelton wheel, the undershot wheel, the water-dropping-on wheel and the water-shooting-out-at wheel. You can use a **rotation sensor** to measure the speeds of rotation in each of these water wheels. The results can help you decide if one design is better than any of the others.

Key stage 4 Measuring using sensors
Idea from Blackwell Modular Science

How do wind vane designs compare?

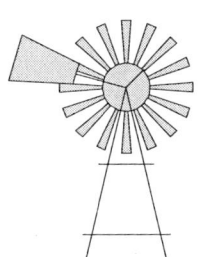

Make up several different wind vane designs. You could use table-tennis balls (anemometer-style); cork and small pieces of card or cork and large pieces of card. Use a **rotation sensor** to measure their speed of rotation over a period of time. You might ask: if you used a windmill as an energy source what problems might you face? Or, why are wind generators used to charge a set of car batteries rather than supply electrical energy directly?

Key stage 3-4 Measuring using sensors

What is renewable energy?

The Internet has much on this area - e.g. visit the web sites of pressure groups.

Electrical energy: domestic use

Compare the energy used by different appliances.

Use an electricity meter (joulemeter) to compare how much energy different electrical appliances use. Use a **spreadsheet** to record the results - for example, you will have columns for meter readings and times. The program can help you calculate the energy used and can also plot a bar graph to compare the appliances you tested.

	A	B	C	D	E
1	How much energy is used by different appliances?				
2	Appliance	Start reading	Final reading	Time	Energy used
3	Hair dryer	1234			
4	Lamp 100 W				
5					

Key stage 4 Handling information using a Spreadsheet. Idea from Salter's Science

How efficient is an electric immersion heater?

Measure the efficiency of an electric immersion heater. Pass current through a low current heater in a beaker of water and measure the temperature rise of water over a specific period of time. Measure also the electrical energy used. A **temperature sensor** will monitor the temperature of the water. A meter or **current sensor** will monitor the flow of electricity. Calculate the efficiency of the heater - by comparing the electrical energy used with the heat gained by the water.
Key stage 4 Measuring using sensors
Idea from Blackwell Modular Science

What's the most appropriate way of getting energy?

The program, **React** (PC - Shell) might be of interest. This demonstrates a broad range of energy producing methods and sets the pupils research tasks.
Key stage 4 Modelling using a Simulation program

How much does it cost to have a bath?

Use a **spreadsheet** to calculate the amount of energy required to heat a bath of water. Similarly you can calculate the amount of energy required for a shower, hand wash, dish wash and clothes wash. You might like to compare the costs of using different energy sources to heat the water. Using a spreadsheet will help with the maths and also allow you to experiment with the figures. See **Electricity** headings, such as Domestic Electricity, for more programs on this topic.

	A	B	C	D	E	F
1	Comparing energy sources					
2	Energy use	Volume required	Working temperature	mst (Energy)	Price	Relative cost
3	Bath	90	54		£0.02	
4	Shower	30			£0.02	
5	Laundry	50			£0.02	
6	Hand wash	5			£0.02	
7	Washing-up	8			£0.02	
8	Cost of energy: look up table					
9	Energy source	Cost / megaJ				
10	Coal	£0.020				
11	Oil	£0.015				
12	Day electricity	£1.700				
13	Night electricity	£0.600				
14	Natural gas	£0.016				

Key stage 3-4 Modelling using a Spreadsheet

Energy: our use of energy

How much energy is used over the world?

For project work and data handling exercises on energy, BP used to produce **The Energy File** (BBC / Nimbus - BP) - a file on energy use across the world but today you can find the info on the web. You used it to collect data on energy use by different countries and decide what factors lie behind the differences. You could enter the data into a spreadsheet, draw graphs and even add them to a word processed report. The **Key** file, **Energy** (All machines - AVP), is a database which you can interrogate using one of the programs in the **Key** family.

	A	B	C	D	E	F
1	How uses the most energy?					
2	Country	Energy per person	Country	Energy per person		
3	Nigeria	6	Mexico	52		
4	Norway	179	Britain	133		
5	Ethiopia	1				
6	Kenya	3				
7	Poland	134				
8	USA	276				
9	Canada	284				
10	Malawi	2				
11	Zambia	11				
12	India	6				
13						
14						

Key stage 3-4 Handling information using a Spreadsheet or Database

Using IT in ... Energy

How has our use of fuels changed over the years?

The use of fuels is a common context for data handling exercises. In textbooks you will find numerous problems on the subject. For example, 'Which fuels where available to us in 1965? Which fuel did we only start using after 1970? How did the total fuel used change over the years between 1960 and 1980?' It is easy to draw graphs and answer such questions using a **spreadsheet** program. At the very least, the graphing facility should save time which could be spent interpreting the data.

> Pupil Worksheet
>
> See the Spreadsheet topic

	A	B	C	D	E	F	G	H
1	How has our use of fuels changed?							
2	Fuel	1960	1965	1970	1975	1980		
3	Town Gas	4	4	3.8	0.5	0		
4	Coal	24	18.5	13	7	5		
5	Natural gas	0	0	1	12	16		
6	Petrol/Oil	14	20	26	24	23		
7	TOTALS	42	42.5	43.8				
8		in 000 mill Therms						
9								
10								
11								
12								
13								
14								
15								
16								

Key stage 3 Handling information using a Spreadsheet
Idea from Blackwell Modular Science / Middle School Science Resources

Electrical energy: generation

How do we generate electrical energy?

Producing Energy (CD-ROM PC / Mac - AVP) uses diagrams and simulations to describe energy production. **Power Control** (PC / Arc - UE) illustrates how electrical energy is generated. It also simulates the supply of electricity to the national grid - responding to peak times of day or national events such as the Olympics. In a sort of game scenario, it highlights the industry's problems in supplying electricity to consumers. **Energy Resources** (CD-ROM PC - AVP) is a tutorial on energy resources and their environmental effects, while **Biosphere (CD-**ROM PC - Education Interactive) is a reasonably detailed encyclopaedia of the environment which gives you the facts about sources of energy and pollution. **Understanding Energy** (CD-ROM PC / Arc - Anglia) examines the different sources of energy, how they were used in the past and how they might be developed in future.

Key stage 4 Modelling using a Simulation program.

We can make electricity using a dynamo on a bicycle. What difference does the speed of the bicycle make?

Use sensors which measure **current** and **potential difference** to study the output from a dynamo. If you also have a **rotation sensor** you may be able to relate the speed of the wheel to the current produced.

Key stage 4 Measuring using sensors Salter's Science

Fuels

Compare the properties and costs of different fuels.

Collect data on fuels - for example on fuel costs and energy content. Then enter the figures into a **spreadsheet**. You can sort the data to find the most expensive fuel or draw a bar graph to find the fuel with the most energy. There is good scope here for doing all sorts of calculations - to find say, the fuel which gives the most energy for the money.

	A	B	C	D	E	F
1	Comparing fuels survey					
2	Fuel	Energy per tonne	Cost per tonne	Cost / energy	Other features	Other
3	Coal					
4	Oil					
5	Paraffin					
6	Coke					
7	Natural gas					

Key stage 3-4 Handling information using a Spreadsheet

Using IT

Section

3

Using IT in ... Energy

Which is the 'best' way to brew up?

Do an experiment to compare the effectiveness of different fuels at heating a quantity of water. Enter the results into a **spreadsheet** and use its calculating features you help you assess the fuel which is 'best'. The spreadsheet can also draw a bar graph to compare the fuels. You can, of course, use temperature sensors to compare the different fuels. For example, you can compare the rate at which they release their energy.

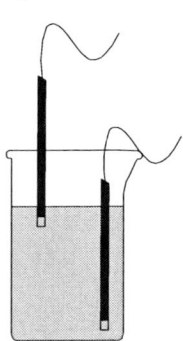

	A	B	C	D	E	F
8	**Comparing fuels experiment**					
9	Fuel	Mass of fuel	Temp of water	Temp after heating	Temp rise	Rise / mass
10	Oil					
11	Firelighter					
12	Alcohol					

Key stage 4 Communicating using a Spreadsheet
Idea from Salter's Science

Which firelighters are the best value?

Burn different brands of firelighter and heat equal volumes of water. Measure the temperature rise. Enter the results into a **spreadsheet** as if you were filling in a results table. Use the program to calculate the energy obtained per gram of fuel. Draw a bar graph of the energy released by each fuel.

	A	B	C	D	E	F
1	**Comparing firelighters for energy**					
2	Fuel	Mass of fuel	Temp of water	Temp after heating	Temp rise	Energy / gram fuel
3	Zippo					
4	Sainsbury					
5	Wood					

Key stage 3-4 Modelling using a Spreadsheet
Idea from Science Scene (Hodder)

Make an advertisement for a fuel.

The energy companies are in competition and they want us to use their product. Use a **word processor** or **graphics program** to prepare one of their advertisements. Mention the fuel's convenience points and advantages to the consumer. To illustrate the work, add diagrams using a scanner. Exercises of this kind can be treated as a good opportunity to get pupils working together and sharing their ideas. Using today's software and printers also encourages pupils to work for a better final product.

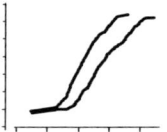

Key stage 3 Communicating using word processing / DTP programs

Heat transfer: liquids

How does heat travel through liquids?

Place two **temperature sensors** in a container of water. Place one at the top and one at the bottom and compare the changes in temperature as you heat the container. Continue monitoring as you let the water cool. You should obtain some interesting evidence for convection - as a graph of temperature against time.

Does water keep getting hotter the more we heat it?

You can use **temperature sensors**, in place of thermometers, to answer questions such as: does water keep getting hotter the more we heat it? Do different amounts of liquid heat up at the same rate? Do large and small cups of coffee cool the same? Does it matter where you put the drink to cool? How can you keep drinks cold for a summer party? How good is a ceramic wine cooler?

Key stage 3-4 Measuring using sensors

Heat transfer: metals

Compare the transfer of heat through different metals.

Use **temperature sensors** to compare the conduction of heat through different metal bars. Not only will you be able to see which metal conducts heat best, you will be able to see the rate of change of temperature too.

Key stage 3-4 Measuring using sensors

Using IT

Section

3

SOFTWARE and CDROM REVIEWS can be found in "SOFTWARE FOR TEACHING SCIENCE" © IT in Science

Using IT in ... Energy

How does the heat flow in a metal bar?

You can 'model' the heat flow in a metal bar, using a **spreadsheet**. You use it to show the temperature in a metal bar as a matrix of numbers. Using formulae, all these numbers are linked to each other. To 'heat' one part of the metal bar you change one of the numbers on the edge of the matrix and immediately 'see' the result ripple through the matrix. There is scope for using this idea in higher level work too - even drawing a surface graph to show the 'hot spots'.

	A	B	C	D	E	F	G	H	I	J	K	L	M	N
1														
2				Temperatures in a										
3				heated block of metal										
4		1	3	4	6	8	12	17	25	42	100			
5		2	5	8	11	16	22	31	45	68	100			
6		3	7	11	16	22	31	42	57	77	100			
7		4	8	13	19	27	36	48	63	81	100			
8		4	9	15	21	30	39	52	66	82	100			
9		5	10	15	22	30	41	53	67	83	100			
10														
11										Heat				
12														
13														

Key stage 4 Modelling using a Spreadsheet
Idea from Information Technology in Science (MEU Cymru)

How fast do copper and aluminium gain heat?

This is an experiment where you heat metal blocks and take their temperatures at regular intervals. If you enter the results into a **spreadsheet** you can plot a graph of temperature against time and work out the specific heat capacity. Similarly, you might use a spreadsheet as an energy calculator. For example, 'if the shc of copper is 380 J/kg/°C how much heat must be given to 2kg copper to raise the temperature by 50°C?'

	A	B	C	D	E	F	G	H	I	J	K	L
1	Heating blocks of metal											
2	Time min	0	1	2	3	4	5	6	7			
3	Metal temperature											
4	Copper	20	30	40	50	60	70					
5	Aluminium	20	24	28	32	36	44					
6												
7												
8												
9												
10												
11												

Heat: radiation

Which gets hotter in the sun?

Temperature sensors allow you to monitor temperature continuously against time. You might use them to compare the temperature of the ground, the grass and say, a metal bench. Or you might use them to compare the temperatures of the 'sea' or the 'beach' during the day. For this second investigation you can use a radiant heat source, a beaker of water as the sea and a beaker of sand as the beach. If you plug the sensors into a data logger you will be able to take readings at the seaside, miles from a computer.

Should you use a black, a white or a shiny material for the roof of the garden shed?

Use **temperature sensors** to measure the temperatures of model sheds sitting under a radiant heat source. Using sensors you can obtain a more 'dynamic' picture of the temperature changes over time. Similarly, you can use temperature sensors to measure the temperature on each surface of a Leslie cube.

Key stage 3-4 Measuring using sensors.
Roof idea from the Oxford Science Programme (OUP)

Heat: insulation

How can you keep some pizza hot for a party?

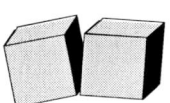
Pupil Worksheet See the Computer Sensors topic

There are endless variations on the 'insulation experiment'. For example, how can you keep some ice-cream cold for a picnic? Or how can we keep our hot water tank hot? Use a pair of **temperature sensors** to compare pizza boxes, thermos flasks, insulating materials and so on.

How long should you leave frozen food to thaw before cooking it?

Freeze say, a sausage with **temperature probes** at different depths inside it. Monitor the change in temperature as the sausage thaws. Or similarly, see how a bread roll freezes - you may even see a depression of the freezing point of water.

Key stage 3-4 Measuring using sensors Pizza idea from the Oxford Science Programme (OUP)

Using IT

Section

3

Using IT in ... Energy

How 'good' is nuclear power?

Write a script for a radio programme about nuclear power using a word processor. Include quotes from the industry and people who are concerned about it.

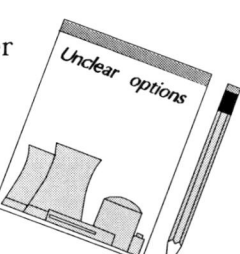

Key stage 3-4 Communicating using a Word processor

How do nuclear reactors work?

Programs showing the working of nuclear power stations and reactors could be useful. For example, **The Nuclear Reactor** (BBC / Arc / Nimbus / PC - AVP) tells about fission, Magnox and AGR reactors.

Key stage 3-4 Modelling using a Simulation program

Saving energy: surveys

How does room temperature vary?

Do a survey to compare temperatures around the room. Use a thermometer or **temperature sensors** to test the inside and outside walls, the floor and the ceiling. You might try to see if the results compare with how warm people feel in different parts of the room. They say that humans give off 100W of heat energy. Use sensors to measure and plot the temperature of the room over a period of time. For example, was there a temperature difference between the start and end of the lesson?

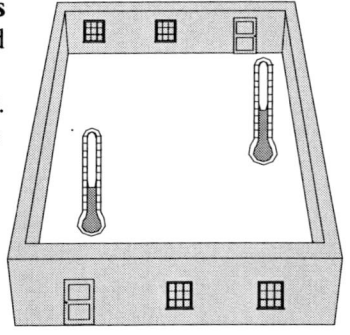

Key stage 3 Measuring using sensors

Does the weather affect how much energy we use?

Investigate how responsive the heating system is. Use **temperature sensors** to compare the indoor and outdoor temperatures over a few days. Alternatively, measure the outside temperature and take readings from the gas / electricity meters daily. Record the results in a **spreadsheet** program. Plot an x-y graph of temperature against energy use to see if we use more energy on cold days.

Key stage 3-4 Measuring using sensors / Modelling using a Spreadsheet

	A	B	C	D	E	F	G	H
1	Does the weather affect how much energy we use?							
2	Date	Inside temp	Outside temp	Electric meter	Gas meter	Electricity used	Gas used	
3	Mon 28.11.94							
4	Fri 2.12.94							
5	Tue 6.12.94							
6								
7	School survey to find out how heating and lighting are being used - I							
8	Room / Activity	Room length	Room Width	Area	Watts / lamp	Total wattage	Watts / area	Light level
9	E201: Lighting							
10	E202: Lighting							
11	E203: Lighting							
12								
13	TOTALS							
14								
15	School survey to find out how heating and lighting are being used - II							
16	Room / Activity	Indoor light	Outdoor light	Daylight factor %	Ext. wall area	Window area	% glazing	
17	E201: Lighting							
18	E202: Lighting							
19	E203: Lighting							

Spreadsheets for energy surveys.

Survey the school to find out how heating and lighting are being used.

Use a thermometer or **temperature sensors** to take readings around the school. Recommended room temperatures list 18°C for a classroom, 16°C for a corridor and 10-13°C for a gym. What energy waste or conservation measures can you find? For example, are outside doors left open?

How much energy does the school waste on lighting in a day? Do a survey of all the rooms in the building. Use a light meter or **light sensor** to find which areas are well lit and which areas have too little light.

Use a **spreadsheet** to record the results as a table *(see above)*. Calculate the amount of energy used and wasted. Suggest how energy could be saved.

Key stage 3-4 Modelling using a Spreadsheet
Idea from the Oxford Science Programme (OUP)

Using IT in ... Energy

Saving energy: double glazing

What can we use to double glaze a window?

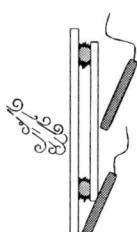

Make a frame of polystyrene tile material and stick this to a window. Attach different coverings to the frame, for example, you can use roasting wrap, glass, cling film, polythene or polypropylene. Place **temperature sensors** in contact with the covering and record the temperature (on a cold day) of the window over a period of time. Which materials prevent heat loss?

Other sensors will also be useful here. A pair of **differential temperature probes** can compare inside and outside temperatures. A **heat flow sensor** can show the **rate** of heat flow through the window. It gives a more dynamic picture of the heat loss.

Does double glazing really work?

Pupil Worksheet
See the Computer Sensors topic

Windows lose a lot of heat and double glazing can be used to reduce this loss. You can investigate this by using two cardboard **model houses**, (ASE) double glazing one and single glazing the other. Heat the houses using small electric lamps. Use **temperature sensors** to measure the 'room' temperatures as the houses get warmer. When the temperatures have reached a steady maximum, allow them to cool. The double glazed house should show its advantage.

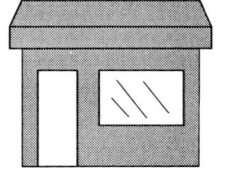

Instead of lamps you can, of course, use beakers of hot water in each house and measure their temperatures. If you have one, a **heat flow sensor** can provide some special information - instead of measuring temperature it can show the **rate** of heat flow through the windows, walls and so on.

Key stage 3 Measuring using sensors

Saving energy: insulation

Which method of home insulation is the most cost-effective?

Pupil Worksheet
See the Spreadsheet topic

There are many ways to keep a house warm and save the energy normally wasted on heating. Some cost more than others. Collect data on the cost of different heat saving methods and even the U values of insulating materials. Enter the data into a **spreadsheet**. Use the program to calculate which methods of insulation are cost-effective.

Key stage 4 Modelling using a Spreadsheet
Idea from the Oxford Science Programme, MEU Cymru and Salter's Science

Make a poster about saving energy ...

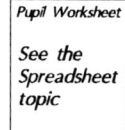

The 'energy' topic offers lots of opportunities for making posters and writing extensively. For example, Make a leaflet for the electricity company explaining how energy can be saved around the home. Make a poster to tell people how to insulate their homes. Write a letter to the manager of a local company explaining how they could use less energy. Write a letter to the architect who is designing your dream home. Specify the work you feel should be done to make it energy-efficient A **word processor** program should be able to help pupils prepare the text in all these examples. A **graphics program** can be used to prepare diagrams explaining say, the ways that hot things cool.

Key stage 3-4. Communicating using a Word processor
Ideas from Kaleidoscope (Heineman) and Blackwell Modular Science

Should we leave the central heating on permanently or switch it off during the day?

It would be nice to be able to create a model of a house, to heat it and then to explore the effectiveness of insulation against the outside temperature. There are a few 'content-free' modelling programs which allow you to set up such a scenario. With them you can say, change the central heating times and so experiment with the model. **Microworlds Project Builder** (PC / Mac - TAG) and **Model Builder** (PC Windows - AU) are intriguing, but you may feel they require too much effort. **Model Builder with Energy Expert** (PC Windows - AU) has ready-made energy models and a touch more accessible. The **Essex Spreadsheets** (PC - Essex) or the **Warwick Spreadsheets** (PC / Mac - Aberdare) each have useful finished examples.

Key stage 3-4 Modelling using a modelling program.
See also School Science Review June 1993

Using IT in ... Forces

Acceleration: gravity

If you dropped a book and a postcard from the same height, which would land first?

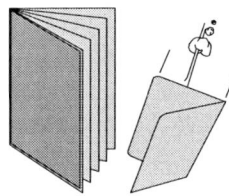

Use timing **light gates** to measure how fast objects fall. You will be impressed with the precision of the measurements. You might measure the average speed or the speeds at different points during the fall.

Using two timing light gates you can measure speeds at two different points and ultimately calculate the acceleration due to gravity.

Your timing software might feature a built in **spreadsheet** - such that the readings you collect can be used to derive other values and plot graphs. In any case, you will find a spreadsheet invaluable for doing calculations such as a= (v-u)/t, in this sort of work. The maths is often a hindrance to pupils' understanding and the spreadsheet can help you make a simple point without distraction.

Key stage 4 Measuring using sensors
Ideas from: The Physical World (Nelson)

Cars

Which is the 'best' car?

Collect data about cars from a magazine. Look for data on engine size, mpg, brake Hp, mass etc. Enter the data into a **spreadsheet** and use the program to calculate the acceleration and see if the relationship between force, mass and acceleration holds true. Choosing the best performer should be fairly easy, and drawing graphs such as mpg against engine size will help this. You can also do a 'Which Magazine' style assessment. You might apply weighting factors to prioritise certain features. For example, if you value the 0-30 time more than the mpg then multiply it by say, 2 to give it a higher weighting.

While we are on the subject see the **Multimedia Motion** II (CD-ROM - PC - Cambridge SM) - a collection of video footage of all kinds of motion. It allows you to perform calculations on people running and cars moving and crashing. This is exceptional. **Four-Stroke Engine** (BBC / Arc / Nimbus - AVP) is a simulation of this difficult to demonstrate engine.

Key stage 4 Handling information using a Spreadsheet
Idea from Salter's Science

What affects our braking distance?

When a car driver stops a car they must react and apply the brakes. The faster they are going, the farther they will travel before stopping the car. A **spreadsheet** can help 'model' this state of affairs. You enter the stopping distances from the highway code. You then can work out the braking distance if you were travelling at say, 150 mph. It's relatively easy to compare graphs of thinking distance against speed with braking distance against speed. You can explore the effect of increasing the reaction time, because of say, tiredness.

	A	B	C	D	E	F	G
1	**Braking distances**						
2	Speed	Driver	Road	Thinking	Braking	Stopping	Car
3	mph	Drunk=2 Alert=1	Wet=2 Dry=1	distance metres	distance metres	distance metres	lengths
4	10	1	1	2.97			
5	20	1	1	5.94			
6	30	1	1	8.91			

Key stage 3-4 Modelling using a Spreadsheet
Idea from Bath Science (Nelson) and Active Science

Distance-Time graphs

Record the journey from your home to your holiday resort.

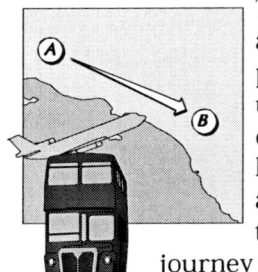

This sort of data is ripe for analysing in a **spreadsheet** program. The spreadsheet can use the data to produce a distance time graph. It can also help you to calculate the average speed for each leg of the journey. For example, the journey might involve a ride to the station, then a train to the main station, you catch another train to the airport, then a plane and a coach. The spreadsheet can use the data to produce a distance time graph.

	A	B	C	D	E	F	G
1	**Distance-time graphs**						
2	Journey to	Start time	End time	Time taken	Distance travelled	Speed mph	Cumulative distance
3	Home	10:10	10:10	0:00	0	0	0
4	Station	10:10	10:30	0:20	0.25	0.75	0.25
5	Main station	10:30	12:00	1:30	70	46.667	70.25
6	Airport	12:00	12:35	0:35	40	68.571	110.25
7	Val d'Isere	12:35	15:30	2:55	700	240	810.25
8	Coach	15:30	18:30	3:00	50	16.667	860.25
9	Hotel	18:30	18:40	0:10	0.25	1.5	860.5
10							
11							
12							
13							
14							
15							
16							
17							
18							
19							

Key stage 3-4 Modelling using a Spreadsheet. Idea from Science Scene (Hodder)

Using IT in ... Forces

Make a distance - time graph.

Using a **distance or motion sensor** it is now possible to make your own distance-time graphs. With the sensor pointing at you, you can walk slowly, walk faster, stand still and hop forwards. You could make a useful 'game' where you show a graph to the class and get them to work out what you were doing at each section of the graph. Or you might challenge them to re-enact the movement for themselves.

Key stage 4 Measuring using sensors
Idea from Science Scene (Hodder)

Flight

How does the load affect the flight time of a 'hot air' balloon?

Results from this kind of investigation are easily recorded and graphed using a **spreadsheet** program. You make a 'balloon' with a carrier, test its flight time with different loads and enter the table of results into the spreadsheet. Using the program's graphing feature, you might plot a bar graph - a series of bars corresponding to the different flights. You might plot an x-y graph - plotting load on one axis and flight time on the other. The pupils might use also a word processor to prepare their report of the investigation. If they do, they should be able to seamlessly incorporate their results and graph into the report.

How does the design of a model aeroplane affect how far it flies?

You can investigate how the position of the wing on a model plane affects how far it can fly. You measure the distance of the wing from the front of the plane and then measure how far it flies. You can build up a table of results in a **spreadsheet** program. The spreadsheet can easily plot an x-y graph of the wing position against the flight. You should find the optimum wing position shown as a peak on the graph.

	A	B	C	D	E	F	G	H
1	Designing a model plane							
2	Distance of wing	1st flight	2nd flight	3rd flight	Average flight			
3	0							
4	1							
5	2							
6	3							

Key stage 3-4 Communicating using a Spreadsheet
Idea from Science Investigations (NCET)

Can you make a good parachute?

Make and test a number of parachutes. As parachute material, you might use paper, tissue, a balloon, a paper plate, polythene sheet, or nylon. You can measure the time the parachute takes to fall. To record the results, enter them into a **spreadsheet**. The program can help you to quickly plot a bar graph to compare the various designs.

Key stage 3 Communicating using a Spreadsheet

How does the size of a sycamore key affect how long it flies?

Collect a large number of sycamore keys. Measure the length / width / weight of each and time how long it takes to fall. Enter the results into a **spreadsheet**. Then get the program to plot x-y graphs of length against fall time, weight against fall time and width against fall time. Is there any pattern in the results?

	A	B	C	D
7	Testing sycamore keys			
8	Key	Mass	Length	Width
9	A			
10	B			
11	C			
12	D			

Key stage 3 Communicating using a Spreadsheet
Idea from Science Investigations (NCET)

Friction

Do all materials produce heat when you rub them?

A **temperature sensor** can provide an on-going measurement of heat output. You could use it to find out if the heat produced depends on the surface and how long you rub it. You might show the effect of oil on the heat produced. It would be interesting to try this with an **infra-red radiation sensor** too.

Key stage 3-4 Measuring using sensors

Using IT in ... Forces

Plan an experiment to compare the brakes on different bikes.

Pupils can use a **word processor** and work with a partner to plan their experiments. You can provide them with a 'template' or pro-forma for planning the experiment. It could have questions such as, "what will you measure? What things must you control? When you have the results, how would you decide which is best?". In this way you provide a structure and focus for their work. Don't overlook the added-value of getting children to plan their work together.

Key stage 4 Communicating using a Word processor
Idea from Science Scene (Hodder)

How does a surface affect how easy it is to slide something along it?

Pull a brick along a wooden floor, Formica table, concrete floor, pile carpet and a nylon carpet tile. Record the results in a **spreadsheet** program. Use the program to sort the results into order. Then plot them on a bar graph to find which surface has the lowest friction. The computer activity here is fairly minimal, yet the time saved in recording and drawing graphs provides some space for pupils to interpret their results. They might even have time to extend their investigation and say, investigate pulling two bricks.

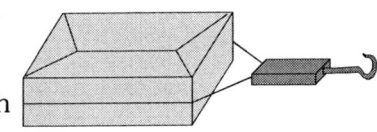

	A	B	C	D	E	F
1	**Friction on different surfaces**					
2	Surface	Distance travelled cm				
3	Glass	60				
4	Vinyl	45				
5	Carpet	20				
6	Wood	40				
7	Alumin	50				
8						

Key stage 3 Modelling using a Spreadsheet
Idea from Folens Copymasters (Folens)

Forces: measuring forces

Measure forces as they change over time.

A **force sensor** is an intriguing device for measuring forces. It increases the scope of the forces you can measure - in particular, it can show you forces changing . The device is essentially a bathroom scale linked to the computer. You might step on and step off it and measure the changing force. You might stand on the scale, take a 3 kg mass from someone and give it back. Or you might walk the plank with one end of the plank on the scale; squash the scale using different muscles or stand on the scale; throw a heavy ball up and then catch it as it falls vertically. Do see the reference below for details.

Key stage 3-4 Measuring using sensors. School Science Review Sept. 1991

Gravity

Write an account of a day in the life of a weightless astronaut.

Use a **word processor** to write up the account. List the sorts of problems the astronaut might have when they are sleeping, walking, keeping fit and eating.

Key stage 4 Communicating using a Word processor. Idea from Science Scene (Hodder)

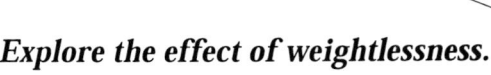

Explore the effect of weightlessness.

You will find many simulations of space walking and gravity-less environments in astronomy programs. **Interactive Physics** (PC / Mac - Logotron) lets you draw a ball on screen and bounce it on different materials or in different gravity - but this is costly. Another title, **Gravity** (Mac / PC - Explorer, TAG) is for advanced level use. It looks at orbits, Kepler's laws, centripetal force, escape velocity, geosynchronous orbits. Programs in this (Explorer) series have earned an above average reputation. **FutureLAB Gravity** (age 13-18 CD-ROM for PC from Nicholl Education) is a photorealistic simulation to use as an electronic blackboard - this is worth seeing. Latest and maybe best is **Crocodile Physics** (for PC from Crocodile Clips)

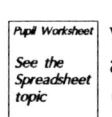

Key stage 3-4 Modelling using a Simulation program

Measure g.

You can use timing **light gates** to measure the acceleration due to gravity.

Key stage 4. See 'Data logging in Practice' (IT in Science)

Using IT in ... Forces

How do the weight readings of things compare in air and in water.

Measure the weights of some objects by suspending them from a force meter. Weigh them hanging in air and then in water. Enter the results into a **spreadsheet** to record the results.

	A	B	C	D	E	F
1	**Weight readings in air and water**					
2		Weight in air g	Weight in water g	Upward push of water	Float or Sink	Mass of water displaced
3	Cone	60 g	0 g	60 g	Float	
4	Bowl	1 g	0 g	1 g	Float	
5	Cube	20 g	0 g	20 g	Float	
6	Stone	80 g	40 g	40 g	Sink	
7	Log	90 g	0 g	90 g	Float	

Key stage 4 Handling information using a Spreadsheet
Idea from Science Scene (Hodder)

Dynamics / conservation of momentum

One Body (Mac /PC - Explorer, TAG) allows you to investigate the motion of a body in a range of inertial systems. You can give the body a charge, add a magnetic field, gravity and frictional forces. The program serves to introduce kinematics and dynamics.

Two Bodies (Mac /PC - Explorer, TAG) allows you to investigate conservation of energy and momentum in two body interactions. "You can add frictional forces, control gravity, modify elasticity and explore Gallilean relativity by doing experiments in moving or accelerating laboratories."

The Cartoon Guide to Physics (Mac/PC for age 15+ on mail order) is a reasonable home or library revision tutorial on forces and motion.
Key stage 3-4, A level. Modelling using a Simulation program

Gravity calculations

Pupil Worksheet

See the Spreadsheet topic

When you need to do calculations, for example to calculate the gravity force (weight = mass x field strength) or potential energy (pe = mgh), use a **spreadsheet** instead of a calculator. The spreadsheet is a surprisingly useful mathematical jotting pad for not just working things out, but also for trying things out - modelling or 'goal seeking' as it is sometimes called.

	A	B	C	D	E
1	**Potential energy calculator**				
2	Example	Mass	Gravity	Height	pe
3	A				
4	B				
5	C				

Key stage 4 Modelling using a Spreadsheet

How does the slope of a hill affect the speed of a free-wheeling vehicle?

Get a vehicle and a ramp and carefully measure how far the vehicle rolls at different angles of the slope. Enter the table of results into a **spreadsheet**. Plot the angle and distance columns as an x-y graph. Can you use the graph to predict the results of angles you have not tried? (See over for the same example using sensors)
Key stage 3 Handling data using a Spreadsheet

Do heavy things roll faster?

Roll a ball, coins, ball bearing, ring, bottle top or marble down a slope and compare how long they take to roll down an unchanging length of slope. Weigh them and enter the table of results into a **spreadsheet**. Sort the things into order of 'roll time'. Plot a bar graph to see if there is a pattern in the results.

	A	B	C	D
1	**How does the angle of the slope affect the speed?**			
2	Slope	Time taken (or distance tra	Rolling experiment	
3	5			
4	10			
5	15			
6	20			
7	25			
8	**How does the mass of the object affect the speed?**			
9	Thing	Time taken (or distance travelled)	Mass	
10	Ball			
11	Coin			
12	Ball bearing			
13	Marble			
14	Bottle top			

Key stage 3-4 Modelling using a Spreadsheet
Idea from Information Technology in Science (MEU Cymru)

Experiments with an air-track.

Using **light gates** and a computer you can explore momentum and many other aspects of motion with an air track. See also Pasco's dynamics track, their trolleys and sensors that measure force, speed and acceleration
Key stage 3-4 Measuring using sensors. See Practical Science with Microcomputers (NCET)

Using IT

Section

3

Using IT in ... Forces

What affects the bounce of a ball?

Collect a set of balls and devise a means of testing how high they bounce. You will need to take a number of readings for each ball and average them. You can then set about finding out whether the mass, the diameter or the material is the deciding factor. A **spreadsheet** can help you record the readings. It can help find out whether the mass of the ball is important, by plotting an x-y graph of average bounce height against mass. Similarly it can plot an x-y graph of average bounce height against diameter. There is plenty more to investigate here: how the height of the drop affects the bounce height or how the height of the drop affects the number of bounces. In all cases, the spreadsheet will be invaluable to help analyse the results.

	A	B	C	D	E	F	G	H
1	What affects the bounce of a ball?							
2	Ball	Try	Try	Try	Try	Average	Mass	Diameter
3	A							
4	B							

Key stage 3-4 Modelling using a Spreadsheet
Idea from Mike Hammond's Handling Data with Databases and Spreadsheets (Hodder)

How does the slope of a hill affect the speed of a free-wheeling vehicle?

Questions such as the above are best answered using **light gates, distance sensors** and other sensors on a computer. Explore the effect of the angle of a ramp: if you double the angle does the speed of the vehicle also double? Or measure the speed of the vehicle at different points down the ramp: why does it change? Or measure the effect of the mass of the vehicle. The scope for experimentation is so vast, that I can best refer readers to **Probing Science** (Data Harvest) and my **Data logging in Practice** (ASE)

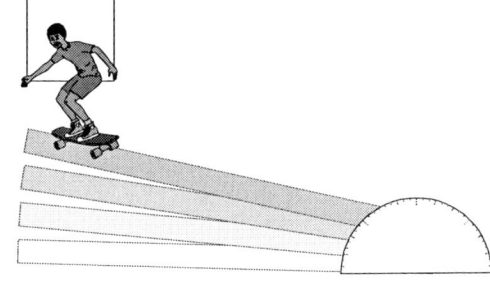

Key stage 3-4 Measuring using sensors
See also School Science Review June 93

Pendulums / Harmonic motion

Explore harmonic motion.

Suspend some weights on a spring and attach it to a position sensor. The sensor will monitor the movements when you disturb the system. You may be able to use a distance or motion sensor to do the same but either way, there is probably no better way to study harmonic motion experimentally. For a high level approach using software see the **Harmonic Motion** simulation (Mac / PC - Explorer, TAG). This allows you to explore the motion of a body from simple one-dimensional oscillations to complex Lissajous figures. You can investigate a variety of damping, driving and constant forces. See also **Harmonic Motion** and **Circular Motion** (PC Visual Products) for a different and impressive approach.

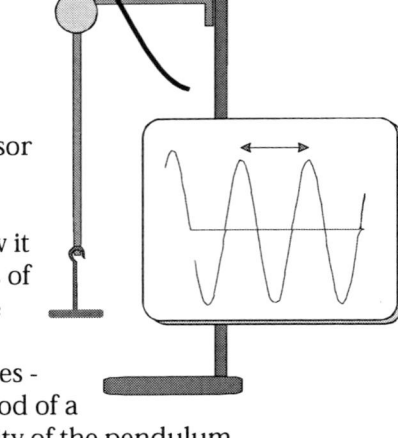

You will also find modelling programs with which you can build your own model of harmonic motion. **Interactive Physics** (PC / Mac - Logotron) does this well as does **Crocodile Physics** (for PC from Crocodile Clips)
Key stage 4 Modelling using a Simulation program

How steady is a pendulum?

You can use IT to help investigate the effects of bob size, pendulum length on the period of a pendulum. In one approach you record the results in a **spreadsheet** - as if you were making a table. You then use the spreadsheet to calculate the period of a single swing or to plot graphs.

A better approach, would be to use a **position sensor** connected to the pendulum. The sensor can record the movement of a pendulum and show it on screen as a series of 'waves'. It's a simple matter to read off values from the waves - finding out the period of a swing and the velocity of the pendulum.

You can also use timing **light gates** to count the number of swings in a given period of time.
Key stage 3-4, A level Handling information using a Spreadsheet/sensors
Spreadsheet idea from Information technology in science (MEU Cymru)

Using IT in ... Forces

Machines and automation

Build an automated machine.

Using 'technology' construction kits you can build a buggy or a robot. You could build a buggy which always moved towards the light or a robot which changed direction when it met an obstacle.

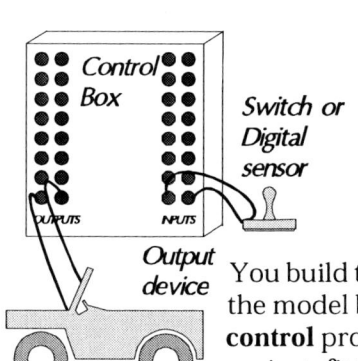

You build the intelligence into the model by using a **computer control** program. These sorts of projects fit in well with microelectronics work.

Incidentally, principles and applications of mechanisms are covered in **Mechanisms** (CD-ROM PC / Arc - AVP).

Key stage 3-4 Measure & Control using control technology

Materials: physical properties

How does the load on a spring affect how much it extends?

This experiment on stretching a spring generates a set of data which could be entered directly into a **spreadsheet program**. The program can plot a graph of spring extension against mass. If you repeat the experiment with other another spring, you can plot both sets of data on a single graph.

	A	B	C	D	E
1	**Extension of a spring**				
2	**Mass**	**Extension**			
3	0				
4	10				
5	20				
6	30				
7	40				
8					
9					

If you have a **position sensor** you can use it to measure the change in extension as the spring is stretched by increasing loads. The sensor software collects and instantly plots the readings. You simply have to tell the computer the size of the mass you use. It should be possible to use a similar setup to compare the elasticity of different materials. For example, you could test a hair or a length of copper wire in place of the spring in the experiment above.

Key stage 3 Modelling using a Spreadsheet Folens / MEU Cymru

What properties would you expect of a vaulting pole?

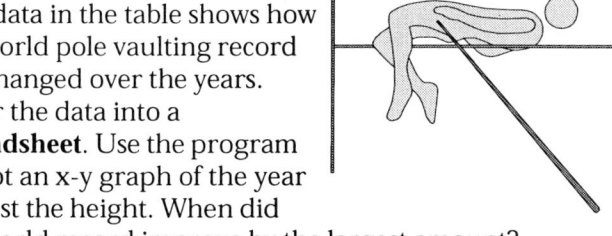

The data in the table shows how the world pole vaulting record has changed over the years. Enter the data into a **spreadsheet**. Use the program to plot an x-y graph of the year against the height. When did the world record improve by the largest amount? What was good about the older pole materials such as bamboo, aluminium and glass fibre? Why do we now use carbon fibre?

	A	B	C	D	E	F	G
1	**Pole vaulting**						
2	**Year**	**World record**					
3	1920	4.2					
4	1930	4.3					
5	1940	4.5					
6	1950	4.6					
7	1960	4.8					
8	1970	5.3					
9	1980	5.7					
10	1990	6.1					

Key stage 3 Handling information using a Spreadsheet
Idea from Science Scene (Hodder)

Can you find the strongest concrete mix?

Make different mixes for concrete and mortar, mould them into bars and allow them to set fully. Test the bars for their breaking strength and use a **spreadsheet** to record the results. The program can quickly and easily plot the results on a 'bar' graph to help you interpret the results.

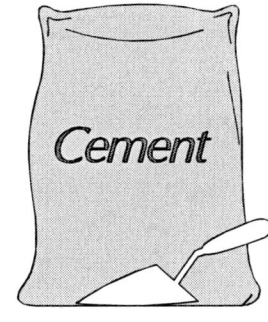

	A	B	C	D
1	**Testing mortar mixes**			
2	**Sand**	**Cement**	**Lime**	**Tested strength**
3	10	1	0.5	
4	8	1	0.5	
5	6	1	0.5	

Key stage 4 Handling information using a Spreadsheet
Idea from The Physical World (Nelson)

Using IT

Section

3

Moments: the see-saw experiment

What is the pattern between the mass and distance on either side of a see-saw?

In this investigation you collect data from various combinations of mass and distance from a fulcrum. You enter the results into a **spreadsheet**. The program can help show the pattern between force and distance by allowing you to quickly calculate which relationship works best i.e. f x d, or f/d or f-d or f+ d.

	A	B	C	D	E	F	G		I	J	K	L	M	N
1	Moments: the see-saw experiment													
2														
3				LEFT SIDE							RIGHT SIDE			
4		No of weights	Dist ance	W x D	W + D	W - D	W / D		No of weights	Dist ance	W x D	W + D	W - D	W / D
5		3	2	6	5	1	1.5		2	3	6	5	-1	0.667
6		3												
7		3												

Key stage 3 Modelling using a Spreadsheet

Power calculations

Measure your power as you lift a weight different distances.

When you need to calculate power (W= f x velocity) enter the data into a **spreadsheet**. You use a formula to calculate power from the force and the velocity. You can usefully play or model with the figures - changing one value and seeing the result on another. This is an easy enough example for beginners.

	A	B	C	D	E	F
1	Work calculator					
2	Job	Height m	Load N	Work J	Time s	Power W
3	A	400	2	800	5	160
4	B					
5	C					
6	D					
7	E					
8	F					
9	Car power calculator					
10	Car	Max speed in mph	in m/s			Power kW
11	X	100	45			36
12	Y	120	54			55

Using IT

Key stage 4 Modelling using a Spreadsheet Understanding Science (John Murray) and Blackwell Modular Science

Section

3

Speed

Does a sprinter run at a steady speed throughout a race?

Pupil Worksheet

See the Spreadsheet topic

I found some data about the 100 metre dash by Ben Johnson at the Olympics*. It showed the time he took to reach various points in a race. It led to some interesting questions: how does his speed change during the race? At what point in the race was he going fastest? The calculations are quite fiddly but using a **spreadsheet**, they are much easier. Using the program allows you to step back from the maths and concentrate on what you are trying to do.

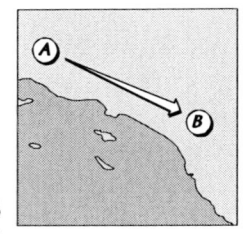

*Idea from *Science Scene 3 (Hodder)*

	A	B	C	D	E
1	Time taken s	Distance run m	Distance travelled m	Time taken for that part of the race	Speed for that portion of the race
2	0.00	0	0	0.00	0.0
3	1.65	10			
4	2.76	20			
5	3.71	30			
6	4.63	40			
7	5.52	50			
8	6.38	60			
9	7.23	70			
10	8.09	80			
11	8.96	90			
12	9.83	100			

Distances between towns

A **spreadsheet** program can be used to make a town to town distance chart. You can use the chart as a 'lookup table' - it can help you calculate how long will it take to travel from A to B if a car has an average speed of 40 mph. Or turning that round you can ask how fast will you need to travel if you need to be at B in 2 hours. You might even work out the total journey time and average speed for a complex itinerary. Essentially, the spreadsheet models the journey of a car.

	A	B	C	D	E	F	G	H	I	J	K	L	M	N	O
1	Town to town travel														
2			Distance	Speed	Time hours	Clock time			Aberdeen	Bristol	Cardiff	Exeter	Hull	Leeds	York
3	Start	Exeter	0	0	0.0	9:00		Bristol	514	~	~	~	~	~	~
4	Finish	Cardiff	121	50	2.4	11:25		Cardiff	534	47	~	~	~	~	~
5	Stopover time				0.6	0:35		Exeter	588	84	121	~	~	~	~
6	Start	Cardiff	0	0	0.0	12:00		Hull	359	231	251	305	~	~	~
7	Finish	Leeds	240	55	4.4	16:22		Leeds	335	220	240	294	61	~	~
8								York	329	225	245	299	37	24	~
9	TOTAL		361					London	546	120	155	200	218	199	212

Key stage 4 Modelling using a Spreadsheet

Using IT in ... Forces

Analyse data from a school athletics event.

You can use a **spreadsheet** and a portable computer to record the results of an athletics event. You can instantly calculate average speeds and also produce a league table.

	A	B	C	D	E
1	**Sports Day**				
2	Event	Name	Distance m	Time s	Average Speed
3	100m	Alison	100	14	
4		Harry			

Key stage 4 Handling information using a Spreadsheet

Study forces and motion

Multimedia Motion (CD-ROM PC - Cambridge SM) has video clips of car crashes, athletes and so on. You can view the video as a series of stills and use mathematical utilities to analyse the motion in detail. This is a unique and enthralling resource to explore motion with.

Forces and their Effects (CD-ROM PC - AVP) covers motion, Newton's laws, pressure, projectiles and more. There are virtual labs where experiments can be tried - lifting this otherwise dry material out of the ordinary. **Picturebase: Physical Processes** (CD-ROM PC / Arc - AVP) is a library of text and photographs about forces, electricity, energy, sound, light and space. **Forces and Motion** (PC CD-ROM for age 12-16 from Granada) has information and investigations on these topics.
Key stage 3-4 Modelling

Structures: bridges

Which bridge design is the 'best'?

Pupil Worksheet

See the Spreadsheet topic

Make a series of bridges to span say, a 15 cm gap. Test how much mass each bridge can take at its centre. You might take this further and find how the span of the bridge affects the strength of one of the bridges. Enter the results into a **spreadsheet** and sort the list of designs into order of their strength. Plot a bar graph of the mass each bridge can handle. A more experienced group could use a word processor to write up their investigation and add their results tables and graphs to their report.

	A	B	C	D	E	F
1	**Designing bridges**					
2	Bridge	Number of Weights take				
3	Flat	0				
4	2 pleats	2				
5	3 pleats	3				
6	Hat-shaped	3				
7	Box	4				
8	Round	4				
9	Triangular	5				
10						
11						
12						

Bridges: Number of weights taken

Key stage 3-4 Modelling using a Spreadsheet
Idea from Science Scene (Hodder)

Projectiles

How does the launch angle of a catapult affect how far a projectile travels?

The sort of data you collect in this exercise can be handled well with a **spreadsheet**. The program can plot the data on a graph with ease. For example you might launch a projectile at a range of different angles and measure how far the projectile travels. You would then plot an x-y graph of distance against angle and see a peak at the optimal launch angle.

	A	B	C	D	E
1	**The best catapult**				
2	Angle	Distance 1	Distance 2	Distance 3	Average distance
3		m	m	m	m
4	0				
5	10				
6	20				
7	30				
8	40				

A more simple alternative would be to compare *different* catapult designs. You enter the results into a spreadsheet in two columns - one for whose design it is, the other for the distances. You then plot a bar graph to compare the catapults.
Key stage 3-4 Handling information using a Spreadsheet
Idea from Science Scene (Hodder)

Advanced forces

The Bradford Advanced Physics series, with titles such as **Mechanics, Thermodynamics & Molecular Physics, Mechanical Oscillations and Waves** should be seen. Lots and lots of models here. (PC CD-ROM - BTL).

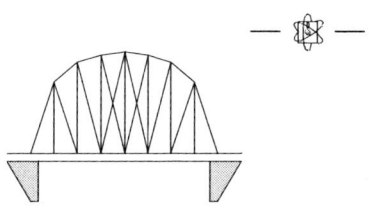

Using IT in ... light and sound

Colour

Mix coloured lights, explore filters and reflection from coloured surfaces.

Mixing Colours (PC- New Media) illustrates several ideas which are difficult to do experimentally.
Light & Sound (PC CD-ROM for age 15+ from BTL) covers the topic dryly with information and experiments.
Key stage 3-4 Modelling using a Simulation program

Who has the best pair of sunglasses?

Use a **light sensor** to test different sunglasses to see which pair are the most effective. You can use home-made glasses made from different coloured film.

How fast do photochromic sunglasses change?

Use a **light sensor** to test some photochromic sunglasses to see how fast they darken and lighten.
Key stage 3-4 Measuring using sensors

Eye: structure and function

Label a diagram of the eye.

Use a **graphics program** to label a diagram of the eye. You can get the picture from the clip-art library of a graphics program or otherwise use a scanner to capture one from a worksheet. The pupils can add labels and even colour it in. On the computer this ought to be a relatively quick exercise with not too much time spent on cosmetic detail. Once the main labels are in place, you can add other labels to show the functions of the various parts. You could also create an exercise where a picture of the eye is assembled from separate parts on the page.
Key stage 3 Communicating using a Graphics program
Idea from Kaleidoscope (Heineman)

Explore the structure of the eye and the functions of its parts.

Computer programs can model long- and short-sightedness and the movement of eye muscles and iris. There's merit in the idea, but too often the programs merely serve as tutorials.
Key stage 3-4 Modelling using a Simulation program

Make a poster for an opticians waiting room.

Use a **word processor** or **graphics program** to prepare a poster about the eye. Let it show how long-sighted and short-sighted people see things differently.
Key stage 3-4 Communicating using word processing / DTP programs
Idea from Kaleidoscope (Heineman)

Compare the parts of the eye with the parts of the camera.

Use a **word processor** to create a table. Tables are a nuisance to create and fill in but on a word processor they are anything but that. Make a 4-column word processor table to compare the eye and the camera, type 'Eye' at the top of one column and 'Camera' at the top of another. Add the parts of the eye and the camera beneath them. Build up the table with facts about the functions of each part.

The eye			
Eye part	**Function in the eye**	**Camera**	**Function in the camera**
Iris	Alter the amount of light entering the lens	Aperture	Alter the amount of light entering the lens
Eye lid			

Key stage 3 Communicating using a Word processor
Idea from Kaleidoscope (Heineman)

Arts & Letters Editor

Using IT

Section
3

Using IT in ... light and sound

Optics

Experiment with lenses and ray boxes.

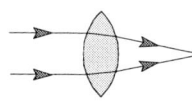

Using ray boxes and lenses, take a series of measurements of angles i and r. Enter the results into a **spreadsheet** instead of a table. The program can calculate sines and focal lengths with ease. You can also use it to 'model' a lens - you might enter the focal length and a value of i and the program will predict the value of r.

	A	B	C	D	E	F
1	Bending rays of light					
2	Angle of incidence	Angle of reflection	i / r	i x r	sin i / sin r	sin i x sin r
3						
4						
5						

One of the examples in the **Warwick** spreadsheet system (PC/Mac - Aberdare) also covers optics. It draws ray diagrams using a spreadsheet. A bit too clever, but still clever. **Crocodile Physics** (for PC from Crocodile Clips) has several excellent teaching tools on the eye, lenses and reflection.

Key stage 4 Modelling using a Spreadsheet
Idea from Understanding Science (John Murray)

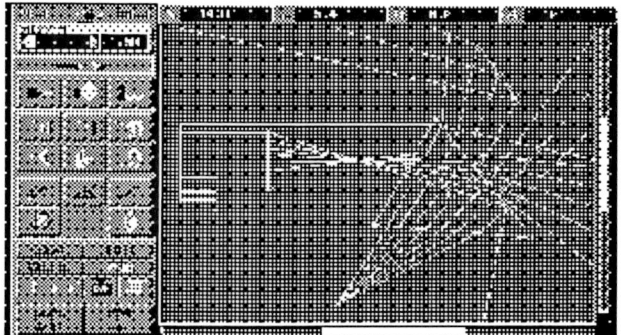

Interference patterns

Study interference patterns

If you direct a laser through a slit, a pattern of the light should emerge the other side of it. You can use a **light sensor** to map the interference pattern - you move the sensor steadily in the same plane as the slit and the pattern of light and dark should appear as a line graph. You might also look at **Diffraction** (Mac/PC Windows - Explorer, TAG). This is a high level look at diffraction, reflection and the interference of light and other waves. It has facilities to vary wavelength and amplitude and to experiment with slits of varying widths and separations. Rated above average but see first the UK title **Multimedia Diffraction** (PC CD-ROM from AVP or Cambridge SM). For lower level work see **Ripple Tank** (for PC from New Media)

Key stage 4 Measuring using sensors

Light and colour

What is found at the ends of the visible spectrum?

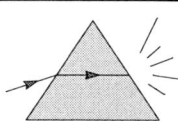

You can detect the **infra-red** and **ultra-violet** at the ends of a spectrum using sensors. For example, using a sensor you will be able to detect infra-red at the end of a spectrum from a prism. Or you could study the change in UV during a day. Without such sensors it's all talk but I can think of few other uses for them to justify the money.

Key stage 4 Measuring using sensors

Which colour clothing would be safest for a cyclist to wear?

Measure the light reflected from different fabrics. Use a **light sensor** to take the readings and so determine the 'best' or brightest clothing to wear.

Key stage 3 Measuring using sensors

Using IT in ... light and sound

Light intensity

Does light fade with distance?

Investigate the inverse square law by using a **light sensor.** The computer can provide a digital read-out of the light level as the distance of the source is varied. Even better, it can take light readings, while you enter distances at the keyboard. It can then plot the relationship on a graph. You should certainly find a pattern but whether you achieve a 'perfect' relationship depends on the characteristics of the light sensor. A **linear light sensor** is more likely to produce a good result.

You might also try to find whether the light reflected from a mirror is greater or less than the incident light level. Try again with two mirrors.

Which light source would be the 'best' to read by?

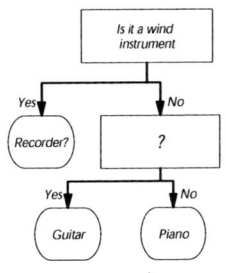

Use a **light sensor** to measure the light levels of sources such as a candle, a torch, a window, a strip lamp and a tungsten lamp. Or use the sensor to compare the light level of different types of candle. In both cases there are plenty of variables to control - making an interesting investigation.

Key stage 3 Measuring using sensors

Do some places in the school have too much or even too little light?

Do a survey around the school to see if energy is being wasted by excessive lighting. Measure the sizes of some school rooms together with their total lighting wattage. If you wish, also use a light sensor with a meter to measure the natural light level. Make a note of rooms where you can you switch off the lights near the window separately and save energy. Enter the results of the survey into a **spreadsheet** program - the program can help you calculate the room areas and the watts / metre. Plot an x-y graph of natural light level or total wattage against room area. Identify action areas for the school.

Key stage 3 Modelling using a Spreadsheet

Musical instruments

Sort out and classify a number of musical instruments.

For your sorting criteria you might pay attention to what makes the sound in a musical instrument or how it is amplified. You might instead focus on what the instrument is made of or whether the pitch is treble or bass. Use a **branching database** program to create a key - it helps you to structure the key and provides an intriguing focus for this work.

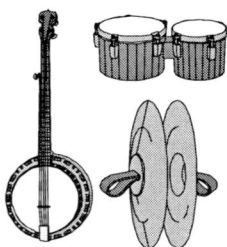

Key stage 3 Handling information using a Database program

Sound travel

How do submarines know where the sea bed is? How do bats avoid collisions?

You can examine the principles of sonar ranging by using a **distance** or **motion sensor**. These sensors work on ultrasound and are a very good way to illustrate sound reflection. The real use of the sensor is for measuring distance, see **Distance-Time graphs** for ideas.

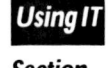

Using IT

Section

3

SOFTWARE and CDROM REVIEWS can be found in "SOFTWARE FOR TEACHING SCIENCE" © IT in Science -

Using IT in ... light and sound

How fast does sound travel?

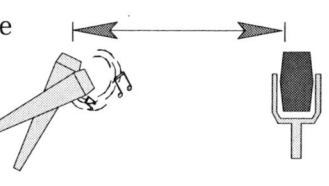

You can actually measure the speed of sound for yourself using a special type of sound sensor called a **sound switch** (Deltronics, Data Harvest and others). You use software which can measure the time between two events: the first event would be to bang together two pieces of metal. This starts the computer timing. The second event occurs when the sound switch picks up the sound. To find the speed of sound, you tell the software what the distance travelled is. The results should be sufficiently impressive to encourage you to investigate how temperature affects the speed of sound, for example, by repeating the experiment out of the window.

Key stage 3-4 Measuring and Modelling using sensors

Sound levels

Investigate sound levels.

A **sound sensor** can be used in numerous investigations. For example, how are the sound levels of musical instruments different? Can you trust your ears to measure? What can sound travel through? Can sound travel through solids and liquids? How can we protect ourselves from unwanted sounds? Which materials make effective sound insulators? Which ear shape is the most directional? Which capture sound the 'best'. See the book **Data logging and Control** (*IT in Science*) for some worked examples.

Key stage 3 Measuring using sensors

Waves

Explore waves.

See **Waves** (Mac /PC - Explorer, TAG). This modelling program allows you to see the motion of waves along a 'string'. You can explore waves in different media and observe velocity changes, reflections and transmission at boundaries. You can also explore superimposition, standing waves, harmonics, damping and driving forces. **Ripple Tank**, a simulation (Mac /PC - New Media) allows you to analyse interference, diffraction, reflection and refraction of waves in a tank. For advanced work see Oscillations and Waves (from www.fable.co.uk)

Multimedia Sound (CD-ROM PC - Cambridge SM) has sound clips which you can analyse using the built-in software. You can also analyse your own sounds. The award for 'best software idea' would go to **PC-Scope** (PC - but UK supplier has gone) which turns the computer into a storage scope and allows you to feed sound into the computer and analyse it. You can do this for real using the fast data loggers from US supplier Pasco.

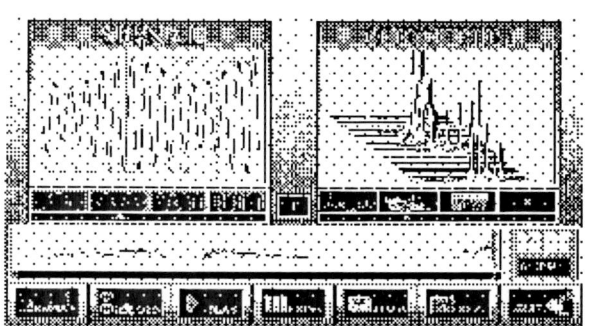

In the same series is **PC-WaveGen**, again for advanced work. This used the computer as a wave generator to good effect:

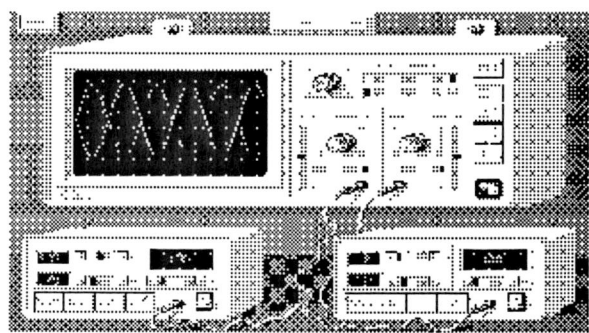

A level Modelling using a simulation program

Using IT in ... earth and space

Explore the phases of the moon, eclipses and the rotation of planets.

A particularly adequate astronomy program is **Orbits** (PC - AVP). This deals with almost every aspect of astronomy going up to advanced level. **Orrery** (Arc - Spacetech) is a simulation of the orbiting planets, **Astro** (Arc / BBC - Topologica) is an easy suite of programs concerning orbits of all kinds. **The Planets Suite** (Nimbus - SPA) is unfortunately weak.

You'll find the most inspired astronomy resources on CD-ROM. By virtue of the fact that they have photographs, film and sounds they can easily impress. **RedShift** (CD-ROM Mac/PC - AVP) is a simulation which shows you, for example, an animated view of the earth from the moon. You can hop around the solar system. It's impressive. **A Field Trip to the Sky** (CD-ROM PC / Mac - TAG) looks at orbits and includes the usual NASA fare. The **Interactive Space Encyclopaedia** (CD-ROM Mac/PC - Andromeda) offers tutorials, a fly around the solar system and lots to look up. Any more? Yes, there's **Eyewitness Encyclopaedia of Space** (CD-ROM PC / Mac - mail) which is quite thorough. Then there's **Earth and Universe** (CD-ROM PC - BTL), **Space Adventure** (CD-ROM PC - Guildsoft), **Space Series: Apollo** (PC / Mac - Optech) and **The view from Earth** (Mac - Optech). **Planetary Taxi** (CD-ROM Mac - Education Interactive) is a solar system game for younger pupils. **Astronomical Explorations** (CDROM - PC / Mac - AVP) has NASA videos while **Distant Suns** (CD-ROM PC / Mac - Education Interactive) is a night sky simulator for astro-enthusiasts. **Solar System Explorer** (PC/Mac CD-ROM for age 11-15 on mail order) is information and simulated voyages. **Nine Worlds** (PC/Mac CD-ROM for age 13-16 on mail order) has a strong documentary style. **The Planets** (PC CD-ROM age 15+ from SCET) is an astronomy offering from Scientific American. **Universe Beyond** (PC CD-ROM on mail order) is a 'for the library' title with nice, black hole fly-throughs and information. **Discovering Astronomy** (PC/Mac CD-ROM for age 15-adult on mail order) includes movies made with Redshift and easier to access.

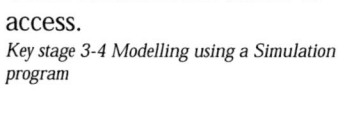

Key stage 3-4 Modelling using a Simulation program

Write a story about the life history of a star.

Pupils can take on the role of a writer for an astronomy magazine. They can use a **word processor** to prepare an account, say, about the life history of a star. If the pupils work together round a screen they'll be able to plan and work together on the story. Those with some IT skills could use a graphics program to add diagrams or scanned images to their account.

Key stage 3 Communicating using a Word processor
Idea from Kaleidoscope (Heineman)

Earth sciences

Rocks

Exploring Earth Sciences (CD-ROM - BTL) covers most of this topic and includes data on rocks, minerals and a satellite atlas of Great Britain.
Picturebase: Rocks, Minerals and Fossils (CDROM PC / Arc - AVP) is a more useful library of text and photographs about minerals, crystals, as well as igneous, sedimentary and metamorphic rocks. **Multimedia Minerals** (CDROM PC - AVP) is a GCSE level look at the characteristics of minerals together with a search tool which can find minerals with similar features. Of great interest to geographers is **Landsat** (CD-ROM PC/Mac/Arc - Timestep) - a service which provides a satellite 'photograph' of an area of your choice together with some stunning analysis software.

Key stage 4 Modelling using a Simulation program

Where do the different rocks come from?

Do a data search to find out where rocks are found. See for example, the **Key** file, **Rocks & Minerals** (All machines - AVP), which works with the **Key** database program.

Key stage 3-4 Handling information using a Database program

Create a key to identify a set of rocks.

Sort out a set of rocks paying special attention to their colour, their hardness, the size of their crystals, whether they are conglomerate and if they contain any useful ingredients. Use a **branching database** program to create an identification key. The program helps you to structure the key and can be the centre of an engaging observation exercise.

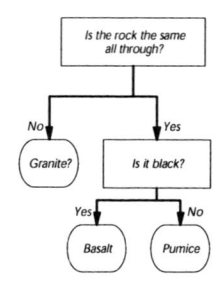

Key stage 3-4 Handling information using a Database program
Idea from Information Technology in Science (MEU Cymru)

SOFTWARE and CDROM REVIEWS can be found in "SOFTWARE FOR TEACHING SCIENCE" © IT in Science

Using IT in ... earth and space

How would you sort out a set of rocks?

You can use a **word processor** to help you sort out a list - and in this example, of rocks. Sorting exercises involve some trial and error and using a word processor facilitates this well. Sort the list using different headings such as hard or soft; large crystals or small crystals.

Key stage 3 Communicating using a Word processor
Idea from the Oxford Science Programme (OUP)

Craters and meteorites

Try this, deliberately misplaced, investigation to find out how the size of a meteorite affects how big a crater it makes. Measure and drop meteorites into a tray of sand. Then place your results in a spreadsheet table under headings such as meteorite diameter, meteorite mass and crater diameter. Draw graphs to find any relationship between the meteorite's mass (or diameter) and how big a crater it makes. You can then repeat this and see how the distance the meteorite falls affects the size of a crater.

Key stage 3 Modelling using a Spreadsheet

Earth

What would life be like if the earth didn't spin?

Use a **word processor** to prepare an account of life on an earth which didn't spin. Consider the effects on the weather, day, night and life. This kind of extended writing exercise presents a good opportunity to use the computer and get the pupils to work collaboratively. They might also use a graphics program or scanner to add any diagrams they need. For example, they might need to include a diagram to show how we get day and night.

Key stage 3 Communicating using a Word processor Graphics
Idea from Science Scene (Hodder)

Make a travel brochure explaining why there are international time differences.

Pupils can use a **word processor** or **graphics program** to prepare posters. They might try to illustrate why it's 8 pm. in Greece when it's only 6 pm. in the UK

Key stage 3 Communicating using a Word processor / modelling with graphics. Idea from Kaleidoscope (Heineman)

Time zones

For a chronosphere showing day, night and time zones see **Small Blue Planet** (CD-ROM PC / Mac - Education Interactive) but only use it if the geographers have already got it as this collection of maps and satellite pictures of the earth is extraordinarily unexciting.

Key stage 4 Modelling using a simulation program

When did the sun rise? When did it set?

Find data on sunrise and sunset and enter it into a **spreadsheet**. You can use the program to calculate the lengths of the day and the night. You can then plot a graph of these against the time of year to see for example how December's days compare to January's. You might also find out when the day starts getting longer.

	A	B	C	D	E	F	G	H	I
1	Sunrise and Sunset								
2	Date	Sun rise	Sun set	Length day	Length night				
3	1.Dec.91	7:45	15:53	8:08	15:52				
4	8.Dec.91	7:54	15:49	7:55	16:05				
5	15.Dec.91	8:01	15:48	7:47	16:13				
6	22.Dec.91	8:06	15:50	7:44	16:16				
7	29.Dec.91	8:08	15:55	7:47	16:13				

It is also interesting to compare your local data with the same in the north or south of the country. Using the spreadsheet you should be able to plot the two sets of data alongside each other.

Computer sensors can also help here. By using a **light sensor** attached to a data logger you can measure the light level over a weekend or even a month. Taking readings over just a few days you'll be able to see if the days are getting longer or shorter. When does a day begin and end anyway? Incidentally, if you have a computer weather station you'll find them better set up for longer term monitoring. Don't be put off by the idea of a long project - just a few days data can be enough.

A computer sky simulator, found in many astronomy programs, will help you find the time of events such as an eclipse.

Key stage 3-4 Modelling using a Spreadsheet. Measuring using sensors. See School Science Review Dec. 92

Using IT in ... earth and space

Tides and the moon

How many tides are there per day? What causes high tides? How long is it from one tide to the next? What's the connection between the phase of moon and the time of high tide?

Find data about the moon, high tides and low tides. Enter the figures into a **spreadsheet** - but code the phase of the moon as a number from 1 to 28. By plotting x-y graphs, for example, you can use the program to help answer the questions above.

	A	B	C	D	E	F	G	H	I
8	The tides and the moon								
9	Date	Moon rise	Moon set	Phase of moon	High tide I	High tide II	Height I	Height II	
10	1.Dec.91	2:33	13:11	4	1:17	14:18	6.3	6.2	
11	2.Dec.91	2:49	13:30	4	2:27	15:17	6.4	6.3	
12	3.Dec.91								
13	4.Dec.91		New moon=1 First q=2						
14	5.Dec.91		Full moon=3 Last q=4						

Key stage 3-4 Modelling using a Spreadsheet. See School Science Review June 93

Moon

Prepare a poster for the classroom wall, about the moon.

Use a **word processor** to write the text for a poster about the moon. Focus on a feature such as the 'seas', the moon's gravity, moon dust, the effect on tides, the phases of the moon or eclipses. Use a **graphics program** to prepare any diagrams you need or better still, scan in some ready-made pictures.

Key stage 3 Communicating using word processing / DTP programs

A moon cut-out exercise

Use a **graphics program** to make a moon cut-out exercise. The idea is to arrange the different shapes of the moon in the sky into the correct order. The program can easily help you draw the set of shapes representing the different phases of the moon. Save the file and get the pupils to arrange the pre-drawn shapes on a plan - showing the positions of the moon, sun and the earth. Once they get the idea they may be able to arrange the shapes, in order, on a 28-day moon diary.

Key stage 3 Communicating using a Graphics program
Idea from Kaleidoscope (Heineman)

Planets

What patterns can you find in data about the planets?

Pupil Worksheet
See the Database topic

Collect together data about the planets, you'll find a data table in most science books. Enter the planet data into a **spreadsheet** or **database** program. These programs help you to analyse the data very easily.

Pupil Worksheet
See the Spreadsheet topic

To start off the pupils can use the 'search' feature of the program to answer questions such as: Which planet is the biggest? Which is the smallest? Which planet has the greatest gravity? Which planet has the shortest day? Which has the longest day? Which is the hottest? Which planet has a year shorter than a day? Which planet takes two years to orbit the sun? Could there be life on Venus?

	A	B	C	D	E	F	G	H	I	J
1	Data on the planets									
2	Planet	Day	Density	Diameter	Distance	Gravity	Mass	Moons	Orbit	Temp
3	Jupiter	9.8	1.34	143000	780	2.6	318	14	12	-150
4	Saturn	10.2	0.7	120000	1430	1.2	95	18	29	-190
5	Uranus	10.8	1.58	50000	2800	1.1	15	15	84	-220
6	Neptune	15.8	2.3	49000	4500	1.4	17	2	165	-240
7	Earth	23.9	5.51	12700	150	1	1	1	1	20
8	Mars	24.6	3.95	6800	228	0.4	0.1	2	1.88	0
9	Pluto	153.6	2	2400	5900	.	0.003	1	248	-240
10	Mercury	1416	5.4	4900	58	0.4	0.05	0	0.24	350
11	Venus	5832	5.25	12100	108	0.1	0.8	0	0.62	480

To compare the planets with each other they can get the program to draw bar graphs of planet diameters, day length, year length, mass, density and surface temperature.

To find patterns between these features, they can get the program to draw x-y graphs of density against mass, number of moons against size, surface temperature against distance from the sun.

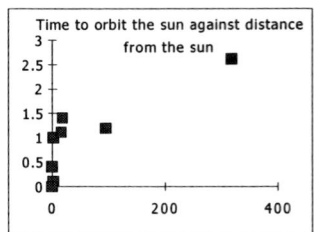

Time to orbit the sun against distance from the sun

You might examine a database on the planets and look for patterns in the data. Or see **Planet Analyser** (New Media) where the hardest part of the work has been done for you.

Key stage 3-4 Handling information using a Spreadsheet or Database. See Enhancing Science (NCET

— —

SOFTWARE and CDROM REVIEWS can be found in "SOFTWARE FOR TEACHING SCIENCE" © IT in Science

Resources and help

Books and things

Data logging in Practice is a companion volume to this. It has ideas and staff training materials for using sensors with pupils aged from 11 to 18 years. ISBN 0 9520257 4 4. From *ASE* or *IT in Science* Published January 1999.

Software for Science Teaching ISBN 0 9520257 5 2 Complements this book by pointing out the ways that software is useful for teaching science to ages 7-18. It aims to help teachers choose software by reviewing, grading and age-banding the titles available. It offers independent advice together with guidelines on the features to look when choosing software and CDROM. The advice is practical and immediately useful - those shopping for science software will find much here to help.

The IT in Science book of Data logging and control is a complementary volume contain a full set of ideas with classroom materials for using sensors with pupils aged from 11 to 18 years. ISBN 0 9520257 1 X. From *ASE* or *IT in Science*

IT in Primary Science - an ideas book, in parallel with this book. Intended for use with pupils aged up to 12 years. ISBN 0 9520257 3 6. From *ASE* or *IT in Science*

Enhancing Science with IT - planning guide, case studies and classroom materials from Becta

Science Online - ideas for using the World Wide Web from Becta

Probing Science - classroom materials for using sensors from Data Harvest.

Probing Science - Electrical Measurements by Roy Barton, the expert on this from Data Harvest

Information Technology in Science - a folder of planning materials and case studies using information technology in science - from MEU Cymru

Model houses - project materials for a domestic energy topic from ASE booksales

Essex Science Spreadsheets - spreadsheet files and ideas for using the Excel spreadsheet - from Essex

Texas Instrument's Interactive and *Mathsoft StudyWorks* for Science (PC CD-ROM on mail order) - two amazing tools for dynamic graphing and handling the algebra in formulae – worth seeing.

Advice and training for science teachers

IT in science specialists - Roger Frost at IT in Science in London (www.rogerfrost.com) and The Science Consortium (www.scienceconsortium.com) provide training and curriculum advice.

National bodies

The national focus for IT in education is Becta in the UK and NTCE in Ireland.

IT Co-ordinators and teachers - ACITT, The National Association for co-ordinators and teachers of IT

Science teaching - ASE, Association for Science Education. Internet: www.ase.org.uk

Resources

Apple software catalogues - from AVP, TAG

Cables and connectors in all forms - Videk Ltd

CD-ROM catalogues - from software suppliers AVP.

General purpose software - from Microsoft, BlackCat, SPA, Flexible software, Logotron, Kudlian Soft, Claris.

Database files - from AVP & Anglia TV. A *Materials datafile* - a Key data file is available from SCSST

Special needs software - from SEMERC and Inclusive Technology.

Warwick Spreadsheet System - ready-made models and tools using the Excel spreadsheet on the PC or Macintosh - from Aberdare.

Data logging and control

Sensors - from Commotion, Deltronics, Data Harvest, Economatics, Griffin & George, Pasco, Scientific and Chemical Supplies.

Data logging software - Examples include *Insight* and *Junior Insight* from Logotron, *Softlab* from Homerton College, *Investigate* from Research Machines. Your sensor supplier may stock these.

Weather stations - from AU and MJP. *Weather satellite stuff* - from Dartcom, MJP, Spacetech and Timestep

Internet places

Updates on this book - Roger Frost's IT in Science site lists software and things worth investigating. Go to www.RogerFrost.com

Information and resources from Becta - go to www.becta.org.uk and find the Virtual Teachers Centre and the National Grid for Learning. See also the ASE site www.ase.org.uk

Using IT

Section

3

Contact details for suppliers

· Updates of this list can be found at www.rogerfrost.com

10 out of 10 Educational Systems, Troydale Mills, Troydale Lane, Leeds, LS28 9LD. Tel: 0113 239 4627

Aberdare Publishing (Warwick Spreadsheet System) 6 Nuthurst Grove, Bentley Heath, Solihull B93 8PD Phone/Fax 01564 773506 Web: members.aol.com/aberdareco

· Acacia Interactive - part of Dorling Kindersley, Web: www.dk.com

· ACITT, The National Association for co-ordinators and teachers of IT, 89 Hutton Avenue, Hartlepool, Cleveland TS26 9PR.

· Aircom Education, PO Box 182, Reigate, Surrey, RH2 0YY. Tel 01737 224434 Fax 01737 222850 Mail: education@aircom.co.uk

· Andromeda Interactive Ltd, 11-15 The Vineyard, Abingdon, Oxford. OX14 3PX. Tel: 01235 529595 Fax: 01235 559122

· Anglia (SCA), PO Box 18, Benfleet, Essex SS7 1AZ. Tel/Fax: 01268 755811 Web: www.anglia.co.uk/education/

· Appian Way Software, Old Co-Operative Buildings, Langley Park, Durham DH7 9XE. Tel: 0191 373 1389

· ASE, Association for Science Education, College Lane, Hatfield. AL10 9AA. Tel: 01707 267411 Fax: 01707 266532 Web: www.ase.org.uk

· Attica Cybernetics, Unit 2, Kings Meadow, Ferry Hinksey Road, Oxford, OX2 0DP. Tel: 01865 791346 Fax: 01865 794561

· AU Enterprises Ltd, 126 Great North Road, Hatfield, Herts, AL9 5JZ. Tel: 01707 266714

· AVP, School Hill Centre, Chepstow, Gwent, NP6 5PH. Telephone: 01291 625439 Web: www.avp.co.uk

· BBC Educational Publishing, PO Box 234, Wetherby, LS23 6YY. Tel: 01937 541001

· BECTA, British Education and Communications Technology Agency, Milburn Hill Road, Science Park, Coventry CV4 7JJ. Telephone: 01203 416994 Fax: 01203 411418 Web: www.becta.org.uk

· BlackCat Educational Software, Second House, Lion House, Bethel Square, Brecon, Powys, LD3 7AY. Tel: 01874 622114 Fax: 01874 611604 Web: www.blackcatsoftware.com

· British Library Publications Office, 41 Russell Square, London, WC1B 3DG, Tel: 0207 412 7535, Fax: 0207 412 7768 Internet: portico.bl.uk

· British Nutrition Foundation, High Holborn House, 52-54 High Holborn, London WC1V 6RQ

· BT Education Services, 81 Newgate Street, London EC1 7AJ. Tel: 0207 356 5677 Fax: 0207 356 5675

· BTL Publishing, Business and Innovation Centre, Angel Way, Listerhills, Bradford, BD7 1BX. Tel: 01274 841320 Fax: 01274 841322 Web: www.bradtech.co.uk

· Cambridge Science Media, 354 Mill Road, Cambridge, CB1 3NN. Tel: 01223 357546. Fax: 01223 573994

· CLEAPSS, School Science Service, Brunel University, Uxbridge, UB8 3PH Tel: 01895 251496 Fax: 01985 814372 Web: www.cleapss.org.uk

· Commotion, Unit 11, Tannery Road, Tonbridge, Kent, TN9 1RF Telephone: 01732 773399

· Computer Concepts, Gaddeston Place, Hemel Hempstead, Herts. HP2 6EX

· Concept Keyboard Company Limited, 9 The Murrils estate, Portchester, Hants, PO16 9RD Tel: 01705 372233 Fax: 01705 372237 Web: www.conceptkey.co.uk

· Creative Curriculum Software, 5 Clover Hill Road, Saville Park, Halifax, HX1 2YG. Telephone: 01422 340524 Fax: 01422 346388 E-mail sales@ccsware.demon.co.uk

· Crocodile Clips Ltd., 11 Randolph Place, Edinburgh, EH3 7TA, Tel: +44 131 226 1511 internet: www.crocodile-clips.com

· Data Harvest, Woburn Lodge, Waterloo Road, Linslade, Leighton Buzzard, Beds., LU7 7NR. Tel: 01525 373666 Fax: 01525 851638 Web: www.data-harvest.co.uk

· Deltronics, Church Road Industrial Estate, Gorslas, Llanelli, Dyfed, SA14 7NF Telephone: 01269 843728 Fax: 01269 845527

· Design Concept, 30 South Oswald Road, Edinburgh, EH9 2HG. Tel/Fax: 0131 668 2000

· Don Johnson Special Needs, 18 Clarendon Court, Calver Road, Winwick Quay, Warrington, WA2 8QP Tel: 01925 241642 Web: www.donjohnson.com

· Dorling Kindersley, 9 Henrietta Street, Covent Garden, London WC2E 8PS Tel: 0207 836 5411 Fax: 0207 836 7570. Web: www.dk.com

· Economatics, Epic House, Darnall Road, Sheffield, S9 5AA. Tel: 0114 2813344 Fax: 0114 2439306

· Education Interactive Ltd (Software catalogue), Hinton House, Hinton, Dorset, BH23 7EA. Telephone: 01425 272235 Fax: 01425 273784

· EMME - Education Multimedia Enterprises, 25 Reddington Road, London, NW3 7QX, Tel: 0207 431 9017 Fax: 0207 431 9025 or contact Interactive Ideas Ltd.

· Essex Science Centre (Essex Spreadsheet Packs), Great Baddow Centre, Duffield Road, Chelmsford, CM3 9SW. Tel: 01245 494291 Fax: 01245 494293

· Europress, Europa House, Adlington Park, Macclesfield, SK10 4NP Tel: 01625 855000 Fax: 01625 855111 Web: www.europress.co.uk

· Fable Multimedia, 116 Weston Park, N8 9PN Tel 0181 374 9008 www.fable.co.uk

· Fable Multimedia, 116 Weston Park, N8 9PN Tel 0208 374 9008 Web www.fable.co.uk

· FlagTower - First Information Group, Knightsbridge House, 197 Knightsbridge, London, SW7 1RB, Tel: 0207 393 3000, Fax: 0207 393 3033

· Flexible Software, PO Box 100, Abingdon, Oxon, OX13 6PQ. Tel: 01865 391148 Fax: 01865 391030 Email: sales@flexible.co.uk

· Focus Multimedia, Lea Hall Enterprise Park, Rugely, WS15 1LH Tel 01889 570156 Fax: 01889 583571 Web: www.focusmm.co.uk

· Future Skill Software, Penrodyn, Pontrhydygroes, Ystrad Meurig, Dyfed, SY25 6DP. Tel: 01974 282428 Web: www.fssc.demon.co.uk

· Georg Thieme Verlag, Rudigerstr. 14 D-70469 Stutgart Tel: 0711-89 31-0 Fax 0711-8931-298

· Global Software Publishing, Meadow Lane, St Ives, Cambs PE17 4LG Tel 01480 496666 Fax 01480 460206 Web: www.gspitd.co.uk

· Granada Learning, Granada Television, Quay Street, Manchester, M60 9EA Tel: 0161 827 2927 Fax: 0161 827 2966 Web: www.granada-learning.com

· Griffin & George, Bishop Meadow Road, Loughborough, Leics., LE11 0RG. Tel: 01509 233344 Fax: 01509 231893

· Guildsoft Ltd, The Software Centre, East Way, Lee Mill Industrial Estate, Ivybridge, PL21 9PE Tel: 01752 895100 Fax: 01752 894833

· Hampshire Microtechnology Centre, The Parkway, 94-96 Wickham Road, Fareham, Hants PO16 7JL Tel: 01329 519111 Fax: 01329 316179

· Helix Educational Software, (Exampro), PO Box 15, Lye, Stourbridge, DY9 7AJ Tel 0184 898969 Fax 01384 426000 Mail www.exampro.co.uk

· Homerton College IT Unit, Cambridge CB2 2PH Telephone: 01223 507161 Fax: 01223 507160

· Inclusive Technology, 2 Castle Street, Castlefield Manchester, M3 4LZ. Tel: 0161 835 3677. Fax: 0161 835 3688 Web: www.inclusive.co.uk

· Interactive Physics, Argenta Research, 35 Arlington Square, N1 7DP Tel: 0171 354 1424

· IT in Science, 7 Sutton Place, London E9 6EH. Tel: / Fax: 0208 986 3526. E-mail: books@rogerfrost.com

· Kudlian Soft, 8 Barrow Road, Kenilworth, Warwickshire, CV8 1EH. Tel/Fax: 01926 851147 E-mail: sales@kudlian.demon.co.uk

· Letts, 9-15 Aldine Street, London W12 8AW. Tel: 0208 740 2270. Internet:www.lettsed.co.uk

· Logotron, 124 Science Park, Milton Road, Cambridge CB4 4ZS Tel: 01223 425558 Web: www.logo.co.uk

Contacts

· Macademic, Trams Ltd, 55-55 Wilton Road, London SW1V 1DE. Tel: 0207 630 6844 Fax: 0207 233 8489

· MAPE (Micros in Primary Education) c/o Yvonne Peers, Newman College, Bartley Green, Birmingham, B32 3NT. Tel 0121 476 1181 x 271.

· Maris Multimedia, 99 Mansell Street, London E1 8AX. Telephone: 0207 488 3029 Tel: 0207 505 1500 Tel: 0207 488 1566 Fax: 0207 702 0534 Web: www.maris.com

· Marshall Cavendish Tel: 0207 734 6710

· MathSoft International, Knightway House, Park Street, Bagshot, GU19 5AQ. Tel: 01276 452299 Fax: 01276 451224. Web: www.mathsoft.com

· Matrix Multimedia Ltd, 10 Hey Street, Bradford, BD7 1DQ Tel/Fax: 01274 730808. Web: www.legend.co.uk/~ matrix/ E-mail: matrix@legend.co.uk

· mc squared, 46b Solent Road, London NW6 1YR tel: 0207 794 7898 Fax 0207 435 4995. Mail 100072.3451@compuserve.com

· McGraw-Hill, Shoppenhangers Lane, Maidenhead, Berkshire, SL6 2QL Tel: 0628 23432 Fax: 01628 770224 Web: www.mcgraw-hill.co.uk

· Media Design Interactive, The Old Hop Kiln, 1 Long Garden Walk, Farnham, Surrey, GU9 7HP. Tel: 01252 737630 Fax: 01252 318775

· Meizner Inc (US software outlet), 4771 Boston Post Road, Pelham, NY 10803. Tel 914 738 6000 Fax 914 738 6068 Mail meizner@aol.com

· MEU Cymru, Gwaelod y Garth Road, Treforest Industrial Estate, Mid Glamorgan, CF37 5US. Tel: 01443 841790

· Microsoft Ltd, Winnersh, Wokingham, Berks, RG11 5TP Tel: 01734 270000 Fax: 01734 270514

· Mindscape Tel: 01444 246333 Web: www.mindscapeuk.com

· MJP-Geopacks, Box 23, St Just, Cornwall, TR19 7JS. Tel: 01736 787808 Fax: 01736 787880

· Modus Project, 1 St James Road, Harpenden, Herts AL5 4NX Tel/Fax: 01582 762297

· Multimedia Textbooks, PO Box 52, Oakham, Rutland LE15 9ZS Tel/Fax 01572 822278 Mail: MMTtheBiz@aol.com

· National Dairy Council, 5-7 John Princes Street, London W1M 0AP. Tel: 0207 499 7822

· NES Arnold, Ludlow Hill Road, West Bridgford, Nottingham, NG2 6HD Tel: 0115 945 2200 Web: www.nesarnold.co.uk

· New Media Press, PO Box 4441, Henley on Thames, Oxon, RG9 3YR Tel: 01491 413999. Fax 01491 574641 Web: www.new-media.co.uk

· New Scientist, Bowker-Saur, Windsor Court, East Grinstead, W Sussex RH19 1XA. Tel: 01342 326972

· Newbyte Educational Software, PO Box 16710, Glascow, G12 9WS Web: www.newbyte.com/uk

· Newman Software, Genners Lane, Bartley Green, Birmingham B32 3NT. Tel: 0121 476 1181

· Nicholl Education Ltd, Block 1, Nortonthorpe Mills, Scissett, Huddersfield HD8 9LA. Tel: 0800 174734 Web: www.nicholl.co.uk

· Oxford Molecular Ltd, Magdalen Centre, Oxford Science Park, Oxford, OX4 4GA.

· Pasco Scientific, c/o Instruments Direct, Windmill Business Centre, Windmill Road, London UB2 4NT. Tel: 0208 560 5678 Fax: 0208 232 8669 Web: www.pasco.com

· Prime Resources, 6 Sunbury Avenue, Jesmond, Newcastle-upon-Tyne, NE2 3HE. Tel 0191 281 1831

· Quickroute, Regent House, Heaton Lane, Stockport SK4 1BS Tel: 0161 476 0202 Fax:0161 476 0505 Web: www.quickroute.co.uk

· Question Mark Computing, Hill House, Highgate Hill, London N19 5NA Tel 0207 263 7575. Fax: 0207 263 7555 Web: www.qmark.co.uk/

· Ransom Publishing, 2 High Street, Watlington, Oxon OX9 5PS. Tel: 01491 613711 Fax:01491 613733 Web: www.ransom.co.uk

· Research Machines Ltd, New Mill House, Milton Park, Abingdon, Oxon, OX14 4BR. Tel: 01235 826000 Fax: 01235 826203 Web: www.rmplc.net

· SCET, 74 Victoria Crescent Road, Glasgow, G12 9JN Sales: 0500 515152 Tel: 0141 337 5000 Fax 0141 337 5050 Web: www.scet.co.uk/ Mail: enquiries@scet.org.uk

· Scientific & Chemical Supplies & Hogg , Carlton House, Livingstone Road, Bilston, WV14 0QZ Tel 01902 402402 01902 402343 Web: www.scichem.co.uk

· Screenactive (Abbey Tutorial Software), 2 The Green, Richmond, Surrey, TW9 1PL Tel 0208 486 1150 Fax: 0208 486 1151 Web: www.trotman.co.uk www.screenactive.co.uk or

· SCSST Materials database, PO Box 92, Wetherby, West Yorkshire, LS23 7TB

· SEMERC, Granada Learning, Granada Television, Quay Street, Manchester, M60 9EA Tel: 0161 827 2927 Fax: 0161 827 2966 Web: www.granada-learning.com

· Shell software, Bankside Business Services, 10 Fleming Road, Newbury, Berks., RG13 2DE.

· Sherston Software, Angel House, Sherston Malmesbury, Wiltshire SN16 0LH. Tel: 01666 840433 Web: www.sherston.com

· SITSS, Bourne House, Radbrook, Shrewsbury, SY3 9BJ. Telephone: 0743 246043 Fax: 0743 368481

· Soft Teach, Sturgess Farmhouse, Longbridge Deverill, Warminster, Wilts BA12 7EA. Fax: 01985 840331 Tel: 01985 840329 www.soft-teach.demon.co.uk

· Softease Ltd, The Old Courthouse, St Peters Church Yard, Derby, DE1 1NN

· SPA, PO Box 59, Tewsbury, GL20 6AB. Tel: 01684 81700 Fax: 01684 81718 www.spasoft.co.uk

· Spacetech Ltd., 21 West Wools, Portland, Dorset DT5 2EA Tel: 01305 822753 Fax: 01305 860483 Web: www.digibase.com/spacetech/

· SSERC, Scottish Schools Equipment Research Centre, St Mary's Building, 23 Holyrood Road, Edinburgh, EH8 8AE. Tel: 0131 558 8180. Fax: 0131 558 8191 Email: sserc@mhie.ac.uk

· Stanley Thornes Publishers, Cheltenham, Glos GL53 1BR. Tel: 01242 228586

· Storm Educational Software, Coachman's Quarters, Digby Road, Sherborne, Dorset DT9 3NN. Tel/Fax: 01935 817699 Internet: ourworld.compuserve.com/homepages/storm_educational

· Swift Test Software, 7 Gowan Avenue, London SW6 6RH. Tel: 0207 731 4108

· TAG Developments, 19 High Street, Gravesend, Kent, DA11 0BA. Tel: 0800 591 262 / 0500 515152 Email: sales@tagdev.co.uk

· The Learning Company, Softkey International, 21 Inner Park Road, London SW19 6ED Tel 0208 246 4000 Fax:0208 246 4029 Web: www.learningco.com

· The National Dairy Council, 5-7 John Princes Street, London W1M 0AP. Telephone: 071 499 7822)

· Timestep Electronics Ltd, Wickhambrook, Newmarket, CB8 8XB. Tel: 01440 820040 Fax: 01440 820281

· Topologica Software, 1 South Harbour, Harbour Village, Penryn, Cornwall, TR10 8LR. Telephone/Fax: 01326 377771 Fax: 01326 376755 www.topolgka.demon.co.uk sales@topolgka.demon.co.uk

· Two-Can, 346 Old Street, EC1V 9NG. Tel 0207 684 4000

· UKAEA, 11 Charles Street, London, SW1Y 4QP

· Understanding electricity, The Electricity Association, 30 Millbank, London, SW1P 4RD. Tel: 0207 344 5768

· Videk Ltd, (cables and connectors), Unit 10, Bowman Trading Estate, Westmoreland Road, London NW9 9RW. Tel: 0208 204 6690

· Visual Products have stopped trading 34 Greenlands Lane, Prestwood, Bucks, HP16 9QU.

· Wayland Publishers, 61 Western Road, Hove, East Sussex BN3 1JD. Tel: 01273 722561. Web: www.emg-ent.com Web site: www.wayland.co.uk

· Whatman International Ltd, St Leonard's Road, Maidstone, Kent ME16 0LS. Tel: 01622 674821 Email: labtech@whatman.co.uk

· WWF, PO Box 963, Slough SL2 3RS. Telephone: 01753 643104

Glossary

Branching Database - a special kind of database. It allows you to build an identification key to sort out a set of animals, plants and so on. A branching database on animals would ask you questions about an animal and eventually it would identify it for you. Using a branching database encourages observation and discussion.

CD-ROM - a computer disc which looks like a compact music disk. However, instead of music the disc stores text, photos, moving images and sounds. You place the disc in a CD-ROM player and see the images on the screen. An incredible amount of information can be stored on one compact disc - an entire encyclopaedia or the equivalent of 700 floppy discs. Pupils can search through it to research and explore a topic. Often there is a measure of interaction and this, of course, is a good starting point for something educational. CD-ROM is like all software - not always wonderful. CD-Interactive is another technology with good potential.

Concept keyboard or overlay keyboard - is an alternative to a button-type keyboard. The keyboard is an A4 or A3 sized tablet which plugs into the computer. Onto you place a sheet of paper, called an overlay. The overlay has words, pictures or even objects on it. When say, a picture is pressed the screen displays some words. This tool can make computing more accessible to pupils - especially younger ones and those with special needs.

Control technology - allows you to control a motorised device, such a fan, with the computer. Using sensors you might arrange for the fan to switch on and off as the temperature changes. Control technology develops problem solving and computer programming skills. It is an aid to understanding how things work. Control technology has a few applications in science teaching.

Database program - a program which lets you store data - such as the data you collect in a survey. You set up a series of headings under which you enter your survey results. Once the data is entered you can search, sort, graph or print the data. You might search a database of people, to find those with dark hair and brown eyes. See also: Database glossary.

Data logging - a method of logging or collecting data from sensors. Strictly speaking, data logging uses devices, called data loggers, which you can take away from the computer and collect data in the field. See the companion to this book called Data logging & Control - which covers this area in depth.

Desk-top publisher - a program to assemble a page with text, borders, boxes and pictures. The text is prepared in a word processor, the pictures in a graphics program. Sometimes a single program does both. A good DTP program and printer can really help produce quite attractive work. Modern word processors have many of the features of DTP programs and are adequate in most cases. A DTP program would need to be really special to justify having one in addition to a good word processor and drawing program.

Digital camera - uses electronic media instead of film. These are the ultimate instant-print cameras and many times more useful. Worth considering if you use cameras as a recording tool in your teaching.

Drawing program - a kind of graphics program where each item on the screen is an object you can scale, move or modify. These are the best choice of program for drawing diagrams. See also Painting program.

Glossary

Graphics programs - these use the computer screen as an electronic canvas. It's very easy to erase mistakes, which helps those of us who can't draw. There are also special features which have no normal comparison - such as painting with striped paint, copying areas, flipping areas upside down or changing their size. Pictures can be pasted into reports, posters and newspapers. Many drawing programs are sufficiently capable to be used instead of desk-top publishing programs.

Multimedia - technology which allows you to experience words, sounds, pictures, animation and/or video when you use the computer. With a modern computer, you can assemble such media yourself to create your own multimedia presentations. A major growth industry with potential for learning about science.

Modelling - a way of representing real-life on the computer. You can experiment with a model and find out how things affect it. A spreadsheet can be used to create a mathematical model of say, how much water we use in a day. You can choose from a whole range of dedicated modelling/simulation programs which make modelling much more accessible.

On-line communications / The Internet - with a computer connected to the phone you can use electronic mail to communicate with schools nationally and internationally. You need to subscribe to an Internet service and use a modem to link the computer to the phone line. The expensive bit is usually the phone bill.

Painting program - a kind of graphics program where you paint on the screen. These are the best choice of program for working with 'art' and photographs. See also *Drawing program*.

Printers - there are many different printer technologies in circulation. The dot-matrix and daisy-wheel printers are pretty-much history now. Ink-jet printers currently offer the best price-performance rating: cheap to buy and with good black print. Colour versions are worth having - they do improve the quality of life. Laser printers are worth aspiring to, everyone should have easy access to one.

Robots - devices which can be programmed to follow directions, draw a trace on the floor or follow a light source. Some of these work independently of the computer, some can be remote-controlled by the computer.

Scanner - an accessory which allows you to capture pictures or photographs for the computer screen. The picture can then be altered, sized and printed alongside the text in a worksheet. An exciting, easy and affordable tool which is well worth a look.

Sensors - there are many sensors - devices which can measure physical quantities such as temperature, light or sound. The measurement can be displayed on a computer screen as a number or graph.

Simulation - a program written to simulate real-life. For example, 'At home in Wattville' is a simulation which shows the use of electricity in the home. You can switch appliances on and off and see the effect on the electricity bill.

Spreadsheet - a program that allows you to handle data in a table on-screen. For example, you could make a spreadsheet to show how much electricity every item in the building used. If you changed the cost of the electricity on the sheet, you would soon see how this would affect how much everything cost to run. The data on screen can also be sorted and graphed - just like a database. Spreadsheets are valuable for handling results from investigations.

Word processor - a program for drafting, editing, printing and improving your written work. Many word processors allow you to change the type style or even add pictures to your work. See also the word processing section.

Using IT

Section

3

Index

SOFTWARE and CDROM REVIEWS can be found in "SOFTWARE FOR TEACHING SCIENCE" © IT in Science